AF810845

MOONPEAK™

PUBLISHERS

RAT CARE JOURNAL

CASE OF EMERGENCY

CARER DETAILS

OWNER NAME	
FULL ADDRESS	
EMAIL & NUMBER	
EMERGENCY CONTACT	
EMERGENCY NUMBER	
DOCTOR DETAILS	

INCIDENT DETAILS

If found please return this

RAT CARE JOURNAL

to:

RAT NAME		DATE
RAT SPECIES		
WEEK STARTING		/ /

RAT AGE		HEALTHCARE RATING	1	2	3	4	5
RAT GENDER							

TIME OF YEAR					FOOD GIVEN	

CLEAN START		
CLEAN END		
RAT SIZE		
RAT WEIGHT		

PET RAT CARE CHECKLIST

☐ CLEAN CAGE	☐ CHANGE TOYS
☐ CHANGE LINING	☐ FOOD/WATER
☐ CHANGE STRAW	☐ CLEAN FAECES

RAT VITAMINS & SUPPLEMENTS

☐ VITAMIN A	☐ VITAMINS C/D	☐ MULTIVITAMINS	☐ OTHER

WEEKLY CARE	MON	TUE	WED	THURS	FRI	SAT	SUN
REFRESH BOWL							
FRESH FOOD							
FRESH WATER							
PLAYING TIME							
EYE CHECK							
EAR CHECK							
CLAW CHECK							
TEETH CHECK							

NOTES & OBSERVATIONS

FINAL ASSESSMENT

OBJECTIVES MET ☐ YES ☐ NO OWNER SIGNED:

RAT NAME		DATE
RAT SPECIES		
WEEK STARTING		/ /

RAT AGE		HEALTHCARE RATING	1	2	3	4	5
RAT GENDER							
TIME OF YEAR		FOOD GIVEN					

CLEAN START		
CLEAN END		
RAT SIZE		
RAT WEIGHT		

PET RAT CARE CHECKLIST

☐ CLEAN CAGE	☐ CHANGE TOYS
☐ CHANGE LINING	☐ FOOD/WATER
☐ CHANGE STRAW	☐ CLEAN FAECES

RAT VITAMINS & SUPPLEMENTS

☐ VITAMIN A	☐ VITAMINS C/D	☐ MULTIVITAMINS	☐ OTHER

WEEKLY CARE	MON	TUE	WED	THURS	FRI	SAT	SUN
REFRESH BOWL							
FRESH FOOD							
FRESH WATER							
PLAYING TIME							
EYE CHECK							
EAR CHECK							
CLAW CHECK							
TEETH CHECK							

NOTES & OBSERVATIONS

FINAL ASSESSMENT

OBJECTIVES MET	☐ YES	☐ NO	OWNER SIGNED:

RAT NAME		DATE
RAT SPECIES		
WEEK STARTING		/ /

RAT AGE			HEALTHCARE RATING	1	2	3	4	5
RAT GENDER								
TIME OF YEAR			FOOD GIVEN					
CLEAN START								
CLEAN END								
RAT SIZE								
RAT WEIGHT								

PET RAT CARE CHECKLIST

- ☐ CLEAN CAGE
- ☐ CHANGE TOYS
- ☐ CHANGE LINING
- ☐ FOOD/WATER
- ☐ CHANGE STRAW
- ☐ CLEAN FAECES

RAT VITAMINS & SUPPLEMENTS

☐ VITAMIN A ☐ VITAMINS C/D ☐ MULTIVITAMINS ☐ OTHER

WEEKLY CARE	MON	TUE	WED	THURS	FRI	SAT	SUN
REFRESH BOWL							
FRESH FOOD							
FRESH WATER							
PLAYING TIME							
EYE CHECK							
EAR CHECK							
CLAW CHECK							
TEETH CHECK							

NOTES & OBSERVATIONS

FINAL ASSESSMENT

OBJECTIVES MET ☐ YES ☐ NO OWNER SIGNED:

RAT NAME		DATE
RAT SPECIES		/ /
WEEK STARTING		

RAT AGE		HEALTHCARE RATING	1	2	3	4	5
RAT GENDER							

TIME OF YEAR					FOOD GIVEN	

CLEAN START		PET RAT CARE CHECKLIST	
CLEAN END		☐ CLEAN CAGE	☐ CHANGE TOYS
RAT SIZE		☐ CHANGE LINING	☐ FOOD/WATER
RAT WEIGHT		☐ CHANGE STRAW	☐ CLEAN FAECES

RAT VITAMINS & SUPPLEMENTS

☐ VITAMIN A ☐ VITAMINS C/D ☐ MULTIVITAMINS ☐ OTHER

WEEKLY CARE	MON	TUE	WED	THURS	FRI	SAT	SUN
REFRESH BOWL							
FRESH FOOD							
FRESH WATER							
PLAYING TIME							
EYE CHECK							
EAR CHECK							
CLAW CHECK							
TEETH CHECK							

NOTES & OBSERVATIONS

FINAL ASSESSMENT

OBJECTIVES MET ☐ YES ☐ NO OWNER SIGNED:

RAT NAME		DATE
RAT SPECIES		
WEEK STARTING		/ /

RAT AGE		HEALTHCARE RATING	1	2	3	4	5
RAT GENDER							

TIME OF YEAR		FOOD GIVEN
CLEAN START		
CLEAN END		
RAT SIZE		
RAT WEIGHT		

PET RAT CARE CHECKLIST

- ☐ CLEAN CAGE
- ☐ CHANGE TOYS
- ☐ CHANGE LINING
- ☐ FOOD/WATER
- ☐ CHANGE STRAW
- ☐ CLEAN FAECES

RAT VITAMINS & SUPPLEMENTS

☐ VITAMIN A ☐ VITAMINS C/D ☐ MULTIVITAMINS ☐ OTHER

WEEKLY CARE	MON	TUE	WED	THURS	FRI	SAT	SUN
REFRESH BOWL							
FRESH FOOD							
FRESH WATER							
PLAYING TIME							
EYE CHECK							
EAR CHECK							
CLAW CHECK							
TEETH CHECK							

NOTES & OBSERVATIONS

FINAL ASSESSMENT

OBJECTIVES MET ☐ YES ☐ NO OWNER SIGNED:

RAT NAME		DATE
RAT SPECIES		
WEEK STARTING		/ /

RAT AGE		HEALTHCARE RATING	1	2	3	4	5
RAT GENDER							
TIME OF YEAR		FOOD GIVEN					
CLEAN START							
CLEAN END							
RAT SIZE							
RAT WEIGHT							

PET RAT CARE CHECKLIST

- ☐ CLEAN CAGE
- ☐ CHANGE TOYS
- ☐ CHANGE LINING
- ☐ FOOD/WATER
- ☐ CHANGE STRAW
- ☐ CLEAN FAECES

RAT VITAMINS & SUPPLEMENTS

☐ VITAMIN A ☐ VITAMINS C/D ☐ MULTIVITAMINS ☐ OTHER

WEEKLY CARE	MON	TUE	WED	THURS	FRI	SAT	SUN
REFRESH BOWL							
FRESH FOOD							
FRESH WATER							
PLAYING TIME							
EYE CHECK							
EAR CHECK							
CLAW CHECK							
TEETH CHECK							

NOTES & OBSERVATIONS

FINAL ASSESSMENT

OBJECTIVES MET ☐ YES ☐ NO OWNER SIGNED:

RAT NAME		DATE
RAT SPECIES		
WEEK STARTING		/ /

RAT AGE		HEALTHCARE RATING	1	2	3	4	5
RAT GENDER							

TIME OF YEAR	FOOD GIVEN

CLEAN START	**PET RAT CARE CHECKLIST**	
CLEAN END	☐ CLEAN CAGE	☐ CHANGE TOYS
RAT SIZE	☐ CHANGE LINING	☐ FOOD/WATER
RAT WEIGHT	☐ CHANGE STRAW	☐ CLEAN FAECES

RAT VITAMINS & SUPPLEMENTS

☐ VITAMIN A ☐ VITAMINS C/D ☐ MULTIVITAMINS ☐ OTHER

WEEKLY CARE	MON	TUE	WED	THURS	FRI	SAT	SUN
REFRESH BOWL							
FRESH FOOD							
FRESH WATER							
PLAYING TIME							
EYE CHECK							
EAR CHECK							
CLAW CHECK							
TEETH CHECK							

NOTES & OBSERVATIONS

FINAL ASSESSMENT

OBJECTIVES MET ☐ YES ☐ NO OWNER SIGNED:

RAT NAME		DATE
RAT SPECIES		
WEEK STARTING		/ /

RAT AGE			HEALTHCARE RATING	1	2	3	4	5
RAT GENDER								
TIME OF YEAR			FOOD GIVEN					
CLEAN START								
CLEAN END								
RAT SIZE								
RAT WEIGHT								

PET RAT CARE CHECKLIST

- ☐ CLEAN CAGE
- ☐ CHANGE TOYS
- ☐ CHANGE LINING
- ☐ FOOD/WATER
- ☐ CHANGE STRAW
- ☐ CLEAN FAECES

RAT VITAMINS & SUPPLEMENTS

☐ VITAMIN A ☐ VITAMINS C/D ☐ MULTIVITAMINS ☐ OTHER

WEEKLY CARE	MON	TUE	WED	THURS	FRI	SAT	SUN
REFRESH BOWL							
FRESH FOOD							
FRESH WATER							
PLAYING TIME							
EYE CHECK							
EAR CHECK							
CLAW CHECK							
TEETH CHECK							

NOTES & OBSERVATIONS

FINAL ASSESSMENT

OBJECTIVES MET ☐ YES ☐ NO OWNER SIGNED:

RAT NAME		DATE
RAT SPECIES		
WEEK STARTING		/ /

RAT AGE		HEALTHCARE RATING	1	2	3	4	5
RAT GENDER							

TIME OF YEAR					FOOD GIVEN	

CLEAN START		
CLEAN END		
RAT SIZE		
RAT WEIGHT		

PET RAT CARE CHECKLIST

☐ CLEAN CAGE	☐ CHANGE TOYS
☐ CHANGE LINING	☐ FOOD/WATER
☐ CHANGE STRAW	☐ CLEAN FAECES

RAT VITAMINS & SUPPLEMENTS

☐ VITAMIN A	☐ VITAMINS C/D	☐ MULTIVITAMINS	☐ OTHER

WEEKLY CARE	MON	TUE	WED	THURS	FRI	SAT	SUN
REFRESH BOWL							
FRESH FOOD							
FRESH WATER							
PLAYING TIME							
EYE CHECK							
EAR CHECK							
CLAW CHECK							
TEETH CHECK							

NOTES & OBSERVATIONS

FINAL ASSESSMENT

OBJECTIVES MET	☐ YES	☐ NO	OWNER SIGNED:

RAT NAME		DATE
RAT SPECIES		
WEEK STARTING		/ /

RAT AGE			HEALTHCARE RATING	1	2	3	4	5
RAT GENDER								
TIME OF YEAR			FOOD GIVEN					
CLEAN START								
CLEAN END								
RAT SIZE								
RAT WEIGHT								

PET RAT CARE CHECKLIST

☐ CLEAN CAGE	☐ CHANGE TOYS
☐ CHANGE LINING	☐ FOOD/WATER
☐ CHANGE STRAW	☐ CLEAN FAECES

RAT VITAMINS & SUPPLEMENTS

☐ VITAMIN A	☐ VITAMINS C/D	☐ MULTIVITAMINS	☐ OTHER

WEEKLY CARE	MON	TUE	WED	THURS	FRI	SAT	SUN
REFRESH BOWL							
FRESH FOOD							
FRESH WATER							
PLAYING TIME							
EYE CHECK							
EAR CHECK							
CLAW CHECK							
TEETH CHECK							

NOTES & OBSERVATIONS

FINAL ASSESSMENT

OBJECTIVES MET	☐ YES	☐ NO	OWNER SIGNED:

RAT NAME		DATE
RAT SPECIES		
WEEK STARTING		/ /

RAT AGE			HEALTHCARE RATING	1	2	3	4	5
RAT GENDER								

TIME OF YEAR						FOOD GIVEN	

CLEAN START		
CLEAN END		
RAT SIZE		
RAT WEIGHT		

PET RAT CARE CHECKLIST

☐ CLEAN CAGE	☐ CHANGE TOYS
☐ CHANGE LINING	☐ FOOD/WATER
☐ CHANGE STRAW	☐ CLEAN FAECES

RAT VITAMINS & SUPPLEMENTS

☐ VITAMIN A	☐ VITAMINS C/D	☐ MULTIVITAMINS	☐ OTHER

WEEKLY CARE	MON	TUE	WED	THURS	FRI	SAT	SUN
REFRESH BOWL							
FRESH FOOD							
FRESH WATER							
PLAYING TIME							
EYE CHECK							
EAR CHECK							
CLAW CHECK							
TEETH CHECK							

NOTES & OBSERVATIONS

FINAL ASSESSMENT

OBJECTIVES MET ☐ YES	☐ NO	OWNER SIGNED:	

RAT NAME			DATE
RAT SPECIES			
WEEK STARTING			/ /

RAT AGE			HEALTHCARE RATING	1	2	3	4	5
RAT GENDER								

TIME OF YEAR					FOOD GIVEN

CLEAN START		
CLEAN END		
RAT SIZE		
RAT WEIGHT		

PET RAT CARE CHECKLIST

☐ CLEAN CAGE	☐ CHANGE TOYS
☐ CHANGE LINING	☐ FOOD/WATER
☐ CHANGE STRAW	☐ CLEAN FAECES

RAT VITAMINS & SUPPLEMENTS

☐ VITAMIN A	☐ VITAMINS C/D	☐ MULTIVITAMINS	☐ OTHER

WEEKLY CARE	MON	TUE	WED	THURS	FRI	SAT	SUN
REFRESH BOWL							
FRESH FOOD							
FRESH WATER							
PLAYING TIME							
EYE CHECK							
EAR CHECK							
CLAW CHECK							
TEETH CHECK							

NOTES & OBSERVATIONS

FINAL ASSESSMENT

OBJECTIVES MET	☐ YES	☐ NO	OWNER SIGNED:

RAT NAME		DATE
RAT SPECIES		
WEEK STARTING		/ /

RAT AGE		HEALTHCARE RATING	1	2	3	4	5
RAT GENDER							

TIME OF YEAR					FOOD GIVEN	

CLEAN START	
CLEAN END	
RAT SIZE	
RAT WEIGHT	

PET RAT CARE CHECKLIST

- ☐ CLEAN CAGE
- ☐ CHANGE TOYS
- ☐ CHANGE LINING
- ☐ FOOD/WATER
- ☐ CHANGE STRAW
- ☐ CLEAN FAECES

RAT VITAMINS & SUPPLEMENTS

☐ VITAMIN A ☐ VITAMINS C/D ☐ MULTIVITAMINS ☐ OTHER

WEEKLY CARE	MON	TUE	WED	THURS	FRI	SAT	SUN
REFRESH BOWL							
FRESH FOOD							
FRESH WATER							
PLAYING TIME							
EYE CHECK							
EAR CHECK							
CLAW CHECK							
TEETH CHECK							

NOTES & OBSERVATIONS

FINAL ASSESSMENT

OBJECTIVES MET ☐ YES ☐ NO OWNER SIGNED:

RAT NAME		DATE
RAT SPECIES		
WEEK STARTING		/ /

RAT AGE		HEALTHCARE RATING	1 | 2 | 3 | 4 | 5
RAT GENDER			
TIME OF YEAR		FOOD GIVEN	
CLEAN START		PET RAT CARE CHECKLIST	
CLEAN END		☐ CLEAN CAGE	☐ CHANGE TOYS
RAT SIZE		☐ CHANGE LINING	☐ FOOD/WATER
RAT WEIGHT		☐ CHANGE STRAW	☐ CLEAN FAECES

RAT VITAMINS & SUPPLEMENTS

☐ VITAMIN A ☐ VITAMINS C/D ☐ MULTIVITAMINS ☐ OTHER

WEEKLY CARE	MON	TUE	WED	THURS	FRI	SAT	SUN
REFRESH BOWL							
FRESH FOOD							
FRESH WATER							
PLAYING TIME							
EYE CHECK							
EAR CHECK							
CLAW CHECK							
TEETH CHECK							

NOTES & OBSERVATIONS

FINAL ASSESSMENT

OBJECTIVES MET ☐ YES ☐ NO OWNER SIGNED:

RAT NAME		DATE
RAT SPECIES		
WEEK STARTING		/ /

RAT AGE			HEALTHCARE RATING	1	2	3	4	5
RAT GENDER								

TIME OF YEAR					FOOD GIVEN	

CLEAN START		
CLEAN END		
RAT SIZE		
RAT WEIGHT		

PET RAT CARE CHECKLIST

- ☐ CLEAN CAGE
- ☐ CHANGE TOYS
- ☐ CHANGE LINING
- ☐ FOOD/WATER
- ☐ CHANGE STRAW
- ☐ CLEAN FAECES

RAT VITAMINS & SUPPLEMENTS

☐ VITAMIN A ☐ VITAMINS C/D ☐ MULTIVITAMINS ☐ OTHER

WEEKLY CARE	MON	TUE	WED	THURS	FRI	SAT	SUN
REFRESH BOWL							
FRESH FOOD							
FRESH WATER							
PLAYING TIME							
EYE CHECK							
EAR CHECK							
CLAW CHECK							
TEETH CHECK							

NOTES & OBSERVATIONS

FINAL ASSESSMENT

OBJECTIVES MET ☐ YES ☐ NO OWNER SIGNED:

RAT NAME		DATE
RAT SPECIES		
WEEK STARTING		/ /

RAT AGE			HEALTHCARE RATING	1	2	3	4	5
RAT GENDER								
TIME OF YEAR			FOOD GIVEN					
CLEAN START								
CLEAN END								
RAT SIZE								
RAT WEIGHT								

PET RAT CARE CHECKLIST

- ☐ CLEAN CAGE
- ☐ CHANGE TOYS
- ☐ CHANGE LINING
- ☐ FOOD/WATER
- ☐ CHANGE STRAW
- ☐ CLEAN FAECES

RAT VITAMINS & SUPPLEMENTS

- ☐ VITAMIN A
- ☐ VITAMINS C/D
- ☐ MULTIVITAMINS
- ☐ OTHER

WEEKLY CARE	MON	TUE	WED	THURS	FRI	SAT	SUN
REFRESH BOWL							
FRESH FOOD							
FRESH WATER							
PLAYING TIME							
EYE CHECK							
EAR CHECK							
CLAW CHECK							
TEETH CHECK							

NOTES & OBSERVATIONS

FINAL ASSESSMENT

OBJECTIVES MET ☐ YES ☐ NO OWNER SIGNED:

RAT NAME		DATE
RAT SPECIES		
WEEK STARTING		/ /

RAT AGE		HEALTHCARE RATING	1	2	3	4	5
RAT GENDER							
TIME OF YEAR		FOOD GIVEN					
CLEAN START		PET RAT CARE CHECKLIST					
CLEAN END		☐ CLEAN CAGE	☐ CHANGE TOYS				
RAT SIZE		☐ CHANGE LINING	☐ FOOD/WATER				
RAT WEIGHT		☐ CHANGE STRAW	☐ CLEAN FAECES				

RAT VITAMINS & SUPPLEMENTS

☐ VITAMIN A ☐ VITAMINS C/D ☐ MULTIVITAMINS ☐ OTHER

WEEKLY CARE	MON	TUE	WED	THURS	FRI	SAT	SUN
REFRESH BOWL							
FRESH FOOD							
FRESH WATER							
PLAYING TIME							
EYE CHECK							
EAR CHECK							
CLAW CHECK							
TEETH CHECK							

NOTES & OBSERVATIONS

FINAL ASSESSMENT

OBJECTIVES MET ☐ YES ☐ NO OWNER SIGNED:

RAT NAME		DATE
RAT SPECIES		
WEEK STARTING		/ /

RAT AGE		HEALTHCARE RATING	1	2	3	4	5
RAT GENDER							
TIME OF YEAR		FOOD GIVEN					
CLEAN START							
CLEAN END							
RAT SIZE							
RAT WEIGHT							

PET RAT CARE CHECKLIST

☐ CLEAN CAGE	☐ CHANGE TOYS
☐ CHANGE LINING	☐ FOOD/WATER
☐ CHANGE STRAW	☐ CLEAN FAECES

RAT VITAMINS & SUPPLEMENTS

☐ VITAMIN A	☐ VITAMINS C/D	☐ MULTIVITAMINS	☐ OTHER

WEEKLY CARE	MON	TUE	WED	THURS	FRI	SAT	SUN
REFRESH BOWL							
FRESH FOOD							
FRESH WATER							
PLAYING TIME							
EYE CHECK							
EAR CHECK							
CLAW CHECK							
TEETH CHECK							

NOTES & OBSERVATIONS

FINAL ASSESSMENT

OBJECTIVES MET ☐ YES ☐ NO OWNER SIGNED:

RAT NAME		DATE
RAT SPECIES		
WEEK STARTING		/ /

RAT AGE			HEALTHCARE RATING	1	2	3	4	5
RAT GENDER								

TIME OF YEAR					FOOD GIVEN

CLEAN START		PET RAT CARE CHECKLIST
CLEAN END		☐ CLEAN CAGE ☐ CHANGE TOYS
RAT SIZE		☐ CHANGE LINING ☐ FOOD/WATER
RAT WEIGHT		☐ CHANGE STRAW ☐ CLEAN FAECES

RAT VITAMINS & SUPPLEMENTS

☐ VITAMIN A ☐ VITAMINS C/D ☐ MULTIVITAMINS ☐ OTHER

WEEKLY CARE	MON	TUE	WED	THURS	FRI	SAT	SUN
REFRESH BOWL							
FRESH FOOD							
FRESH WATER							
PLAYING TIME							
EYE CHECK							
EAR CHECK							
CLAW CHECK							
TEETH CHECK							

NOTES & OBSERVATIONS

FINAL ASSESSMENT

OBJECTIVES MET ☐ YES ☐ NO OWNER SIGNED:

RAT NAME		DATE
RAT SPECIES		
WEEK STARTING		/ /

RAT AGE		HEALTHCARE RATING	1	2	3	4	5
RAT GENDER							

TIME OF YEAR		FOOD GIVEN	

CLEAN START		
CLEAN END		
RAT SIZE		
RAT WEIGHT		

PET RAT CARE CHECKLIST

☐ CLEAN CAGE	☐ CHANGE TOYS
☐ CHANGE LINING	☐ FOOD/WATER
☐ CHANGE STRAW	☐ CLEAN FAECES

RAT VITAMINS & SUPPLEMENTS

☐ VITAMIN A ☐ VITAMINS C/D ☐ MULTIVITAMINS ☐ OTHER

WEEKLY CARE	MON	TUE	WED	THURS	FRI	SAT	SUN
REFRESH BOWL							
FRESH FOOD							
FRESH WATER							
PLAYING TIME							
EYE CHECK							
EAR CHECK							
CLAW CHECK							
TEETH CHECK							

NOTES & OBSERVATIONS

FINAL ASSESSMENT

OBJECTIVES MET ☐ YES ☐ NO OWNER SIGNED:

RAT NAME		DATE
RAT SPECIES		
WEEK STARTING		/ /

RAT AGE		HEALTHCARE RATING	1	2	3	4	5
RAT GENDER							

TIME OF YEAR					FOOD GIVEN	

CLEAN START		PET RAT CARE CHECKLIST	
CLEAN END		☐ CLEAN CAGE	☐ CHANGE TOYS
RAT SIZE		☐ CHANGE LINING	☐ FOOD/WATER
RAT WEIGHT		☐ CHANGE STRAW	☐ CLEAN FAECES

RAT VITAMINS & SUPPLEMENTS

☐ VITAMIN A ☐ VITAMINS C/D ☐ MULTIVITAMINS ☐ OTHER

WEEKLY CARE	MON	TUE	WED	THURS	FRI	SAT	SUN
REFRESH BOWL							
FRESH FOOD							
FRESH WATER							
PLAYING TIME							
EYE CHECK							
EAR CHECK							
CLAW CHECK							
TEETH CHECK							

NOTES & OBSERVATIONS

FINAL ASSESSMENT

OBJECTIVES MET ☐ YES ☐ NO OWNER SIGNED:

RAT NAME		DATE
RAT SPECIES		
WEEK STARTING		/ /

RAT AGE		HEALTHCARE RATING	1	2	3	4	5
RAT GENDER							

TIME OF YEAR		FOOD GIVEN	

CLEAN START		PET RAT CARE CHECKLIST	
CLEAN END		☐ CLEAN CAGE	☐ CHANGE TOYS
RAT SIZE		☐ CHANGE LINING	☐ FOOD/WATER
RAT WEIGHT		☐ CHANGE STRAW	☐ CLEAN FAECES

RAT VITAMINS & SUPPLEMENTS

☐ VITAMIN A ☐ VITAMINS C/D ☐ MULTIVITAMINS ☐ OTHER

WEEKLY CARE	MON	TUE	WED	THURS	FRI	SAT	SUN
REFRESH BOWL							
FRESH FOOD							
FRESH WATER							
PLAYING TIME							
EYE CHECK							
EAR CHECK							
CLAW CHECK							
TEETH CHECK							

NOTES & OBSERVATIONS

FINAL ASSESSMENT

OBJECTIVES MET ☐ YES ☐ NO OWNER SIGNED:

RAT NAME		DATE
RAT SPECIES		
WEEK STARTING		/ /

RAT AGE			HEALTHCARE RATING	1	2	3	4	5
RAT GENDER								

TIME OF YEAR					FOOD GIVEN	

CLEAN START		
CLEAN END		
RAT SIZE		
RAT WEIGHT		

PET RAT CARE CHECKLIST

☐ CLEAN CAGE	☐ CHANGE TOYS
☐ CHANGE LINING	☐ FOOD/WATER
☐ CHANGE STRAW	☐ CLEAN FAECES

RAT VITAMINS & SUPPLEMENTS

☐ VITAMIN A ☐ VITAMINS C/D ☐ MULTIVITAMINS ☐ OTHER

WEEKLY CARE	MON	TUE	WED	THURS	FRI	SAT	SUN
REFRESH BOWL							
FRESH FOOD							
FRESH WATER							
PLAYING TIME							
EYE CHECK							
EAR CHECK							
CLAW CHECK							
TEETH CHECK							

NOTES & OBSERVATIONS

FINAL ASSESSMENT

OBJECTIVES MET ☐ YES ☐ NO OWNER SIGNED:

RAT NAME		DATE
RAT SPECIES		
WEEK STARTING		/ /

RAT AGE		HEALTHCARE RATING	1	2	3	4	5
RAT GENDER							

TIME OF YEAR					FOOD GIVEN	

CLEAN START	
CLEAN END	
RAT SIZE	
RAT WEIGHT	

PET RAT CARE CHECKLIST

☐ CLEAN CAGE	☐ CHANGE TOYS
☐ CHANGE LINING	☐ FOOD/WATER
☐ CHANGE STRAW	☐ CLEAN FAECES

RAT VITAMINS & SUPPLEMENTS

☐ VITAMIN A	☐ VITAMINS C/D	☐ MULTIVITAMINS	☐ OTHER

WEEKLY CARE	MON	TUE	WED	THURS	FRI	SAT	SUN
REFRESH BOWL							
FRESH FOOD							
FRESH WATER							
PLAYING TIME							
EYE CHECK							
EAR CHECK							
CLAW CHECK							
TEETH CHECK							

NOTES & OBSERVATIONS

FINAL ASSESSMENT

OBJECTIVES MET ☐ YES	☐ NO	OWNER SIGNED:	

RAT NAME		DATE
RAT SPECIES		
WEEK STARTING		/ /

RAT AGE		HEALTHCARE RATING	1	2	3	4	5
RAT GENDER							

TIME OF YEAR		FOOD GIVEN

CLEAN START		**PET RAT CARE CHECKLIST**
CLEAN END		☐ CLEAN CAGE ☐ CHANGE TOYS
RAT SIZE		☐ CHANGE LINING ☐ FOOD/WATER
RAT WEIGHT		☐ CHANGE STRAW ☐ CLEAN FAECES

RAT VITAMINS & SUPPLEMENTS

☐ VITAMIN A ☐ VITAMINS C/D ☐ MULTIVITAMINS ☐ OTHER

WEEKLY CARE	MON	TUE	WED	THURS	FRI	SAT	SUN
REFRESH BOWL							
FRESH FOOD							
FRESH WATER							
PLAYING TIME							
EYE CHECK							
EAR CHECK							
CLAW CHECK							
TEETH CHECK							

NOTES & OBSERVATIONS

FINAL ASSESSMENT

OBJECTIVES MET ☐ YES ☐ NO OWNER SIGNED:

RAT NAME		DATE
RAT SPECIES		
WEEK STARTING		/ /

RAT AGE		HEALTHCARE RATING	1	2	3	4	5
RAT GENDER							
TIME OF YEAR		FOOD GIVEN					
CLEAN START							
CLEAN END							
RAT SIZE							
RAT WEIGHT							

PET RAT CARE CHECKLIST

- ☐ CLEAN CAGE
- ☐ CHANGE TOYS
- ☐ CHANGE LINING
- ☐ FOOD/WATER
- ☐ CHANGE STRAW
- ☐ CLEAN FAECES

RAT VITAMINS & SUPPLEMENTS

- ☐ VITAMIN A
- ☐ VITAMINS C/D
- ☐ MULTIVITAMINS
- ☐ OTHER

WEEKLY CARE	MON	TUE	WED	THURS	FRI	SAT	SUN
REFRESH BOWL							
FRESH FOOD							
FRESH WATER							
PLAYING TIME							
EYE CHECK							
EAR CHECK							
CLAW CHECK							
TEETH CHECK							

NOTES & OBSERVATIONS

FINAL ASSESSMENT

OBJECTIVES MET ☐ YES ☐ NO OWNER SIGNED:

RAT NAME		DATE
RAT SPECIES		
WEEK STARTING		/ /

RAT AGE		HEALTHCARE RATING	1	2	3	4	5
RAT GENDER							

TIME OF YEAR					FOOD GIVEN

CLEAN START	
CLEAN END	
RAT SIZE	
RAT WEIGHT	

PET RAT CARE CHECKLIST

☐ CLEAN CAGE	☐ CHANGE TOYS
☐ CHANGE LINING	☐ FOOD/WATER
☐ CHANGE STRAW	☐ CLEAN FAECES

RAT VITAMINS & SUPPLEMENTS

☐ VITAMIN A	☐ VITAMINS C/D	☐ MULTIVITAMINS	☐ OTHER

WEEKLY CARE	MON	TUE	WED	THURS	FRI	SAT	SUN
REFRESH BOWL							
FRESH FOOD							
FRESH WATER							
PLAYING TIME							
EYE CHECK							
EAR CHECK							
CLAW CHECK							
TEETH CHECK							

NOTES & OBSERVATIONS

FINAL ASSESSMENT

OBJECTIVES MET ☐ YES ☐ NO OWNER SIGNED:

RAT NAME		DATE
RAT SPECIES		
WEEK STARTING		/ /

RAT AGE			HEALTHCARE RATING	1	2	3	4	5
RAT GENDER								

TIME OF YEAR						FOOD GIVEN	

CLEAN START		
CLEAN END		
RAT SIZE		
RAT WEIGHT		

PET RAT CARE CHECKLIST

☐ CLEAN CAGE	☐ CHANGE TOYS
☐ CHANGE LINING	☐ FOOD/WATER
☐ CHANGE STRAW	☐ CLEAN FAECES

RAT VITAMINS & SUPPLEMENTS

☐ VITAMIN A	☐ VITAMINS C/D	☐ MULTIVITAMINS	☐ OTHER

WEEKLY CARE	MON	TUE	WED	THURS	FRI	SAT	SUN
REFRESH BOWL							
FRESH FOOD							
FRESH WATER							
PLAYING TIME							
EYE CHECK							
EAR CHECK							
CLAW CHECK							
TEETH CHECK							

NOTES & OBSERVATIONS

FINAL ASSESSMENT

OBJECTIVES MET	☐ YES	☐ NO	OWNER SIGNED:

RAT NAME		DATE
RAT SPECIES		
WEEK STARTING		/ /

RAT AGE			HEALTHCARE RATING	1	2	3	4	5
RAT GENDER								

TIME OF YEAR					FOOD GIVEN	

CLEAN START			PET RAT CARE CHECKLIST	
CLEAN END			☐ CLEAN CAGE	☐ CHANGE TOYS
RAT SIZE			☐ CHANGE LINING	☐ FOOD/WATER
RAT WEIGHT			☐ CHANGE STRAW	☐ CLEAN FAECES

RAT VITAMINS & SUPPLEMENTS

☐ VITAMIN A ☐ VITAMINS C/D ☐ MULTIVITAMINS ☐ OTHER

WEEKLY CARE	MON	TUE	WED	THURS	FRI	SAT	SUN
REFRESH BOWL							
FRESH FOOD							
FRESH WATER							
PLAYING TIME							
EYE CHECK							
EAR CHECK							
CLAW CHECK							
TEETH CHECK							

NOTES & OBSERVATIONS

FINAL ASSESSMENT

OBJECTIVES MET ☐ YES ☐ NO OWNER SIGNED:

RAT NAME		DATE
RAT SPECIES		
WEEK STARTING		/ /

RAT AGE		HEALTHCARE RATING	1	2	3	4	5
RAT GENDER							

TIME OF YEAR					FOOD GIVEN	

CLEAN START	
CLEAN END	
RAT SIZE	
RAT WEIGHT	

PET RAT CARE CHECKLIST

☐ CLEAN CAGE	☐ CHANGE TOYS
☐ CHANGE LINING	☐ FOOD/WATER
☐ CHANGE STRAW	☐ CLEAN FAECES

RAT VITAMINS & SUPPLEMENTS

☐ VITAMIN A	☐ VITAMINS C/D	☐ MULTIVITAMINS	☐ OTHER

WEEKLY CARE	MON	TUE	WED	THURS	FRI	SAT	SUN
REFRESH BOWL							
FRESH FOOD							
FRESH WATER							
PLAYING TIME							
EYE CHECK							
EAR CHECK							
CLAW CHECK							
TEETH CHECK							

NOTES & OBSERVATIONS

FINAL ASSESSMENT

OBJECTIVES MET	☐ YES	☐ NO	OWNER SIGNED:

RAT NAME		DATE
RAT SPECIES		
WEEK STARTING		/ /

RAT AGE		HEALTHCARE RATING	1	2	3	4	5
RAT GENDER							
TIME OF YEAR		FOOD GIVEN					

CLEAN START		PET RAT CARE CHECKLIST	
CLEAN END		☐ CLEAN CAGE	☐ CHANGE TOYS
RAT SIZE		☐ CHANGE LINING	☐ FOOD/WATER
RAT WEIGHT		☐ CHANGE STRAW	☐ CLEAN FAECES

RAT VITAMINS & SUPPLEMENTS

☐ VITAMIN A ☐ VITAMINS C/D ☐ MULTIVITAMINS ☐ OTHER

WEEKLY CARE	MON	TUE	WED	THURS	FRI	SAT	SUN
REFRESH BOWL							
FRESH FOOD							
FRESH WATER							
PLAYING TIME							
EYE CHECK							
EAR CHECK							
CLAW CHECK							
TEETH CHECK							

NOTES & OBSERVATIONS

FINAL ASSESSMENT

OBJECTIVES MET ☐ YES ☐ NO OWNER SIGNED:

RAT NAME		DATE
RAT SPECIES		
WEEK STARTING		/ /

RAT AGE		HEALTHCARE RATING 1 2 3 4 5
RAT GENDER		
TIME OF YEAR		FOOD GIVEN
CLEAN START		
CLEAN END		
RAT SIZE		
RAT WEIGHT		

PET RAT CARE CHECKLIST

- ☐ CLEAN CAGE
- ☐ CHANGE TOYS
- ☐ CHANGE LINING
- ☐ FOOD/WATER
- ☐ CHANGE STRAW
- ☐ CLEAN FAECES

RAT VITAMINS & SUPPLEMENTS

☐ VITAMIN A ☐ VITAMINS C/D ☐ MULTIVITAMINS ☐ OTHER

WEEKLY CARE	MON	TUE	WED	THURS	FRI	SAT	SUN
REFRESH BOWL							
FRESH FOOD							
FRESH WATER							
PLAYING TIME							
EYE CHECK							
EAR CHECK							
CLAW CHECK							
TEETH CHECK							

NOTES & OBSERVATIONS

FINAL ASSESSMENT

OBJECTIVES MET ☐ YES ☐ NO OWNER SIGNED:

RAT NAME		DATE
RAT SPECIES		
WEEK STARTING		/ /

RAT AGE			HEALTHCARE RATING	1	2	3	4	5
RAT GENDER								

TIME OF YEAR						FOOD GIVEN	

CLEAN START		
CLEAN END		
RAT SIZE		
RAT WEIGHT		

PET RAT CARE CHECKLIST

☐ CLEAN CAGE	☐ CHANGE TOYS
☐ CHANGE LINING	☐ FOOD/WATER
☐ CHANGE STRAW	☐ CLEAN FAECES

RAT VITAMINS & SUPPLEMENTS

☐ VITAMIN A	☐ VITAMINS C/D	☐ MULTIVITAMINS	☐ OTHER

WEEKLY CARE	MON	TUE	WED	THURS	FRI	SAT	SUN
REFRESH BOWL							
FRESH FOOD							
FRESH WATER							
PLAYING TIME							
EYE CHECK							
EAR CHECK							
CLAW CHECK							
TEETH CHECK							

NOTES & OBSERVATIONS

FINAL ASSESSMENT

OBJECTIVES MET	☐ YES	☐ NO	OWNER SIGNED:

RAT NAME		DATE
RAT SPECIES		
WEEK STARTING		/ /

RAT AGE					HEALTHCARE RATING	1	2	3	4	5
RAT GENDER										
TIME OF YEAR					FOOD GIVEN					
CLEAN START					PET RAT CARE CHECKLIST					
CLEAN END					☐ CLEAN CAGE	☐ CHANGE TOYS				
RAT SIZE					☐ CHANGE LINING	☐ FOOD/WATER				
RAT WEIGHT					☐ CHANGE STRAW	☐ CLEAN FAECES				

RAT VITAMINS & SUPPLEMENTS

☐ VITAMIN A ☐ VITAMINS C/D ☐ MULTIVITAMINS ☐ OTHER

WEEKLY CARE	MON	TUE	WED	THURS	FRI	SAT	SUN
REFRESH BOWL							
FRESH FOOD							
FRESH WATER							
PLAYING TIME							
EYE CHECK							
EAR CHECK							
CLAW CHECK							
TEETH CHECK							

NOTES & OBSERVATIONS

FINAL ASSESSMENT

OBJECTIVES MET ☐ YES ☐ NO OWNER SIGNED:

RAT NAME		DATE
RAT SPECIES		
WEEK STARTING		/ /

RAT AGE		HEALTHCARE RATING	1	2	3	4	5
RAT GENDER							

TIME OF YEAR					FOOD GIVEN	

CLEAN START	
CLEAN END	
RAT SIZE	
RAT WEIGHT	

PET RAT CARE CHECKLIST

☐ CLEAN CAGE	☐ CHANGE TOYS
☐ CHANGE LINING	☐ FOOD/WATER
☐ CHANGE STRAW	☐ CLEAN FAECES

RAT VITAMINS & SUPPLEMENTS

☐ VITAMIN A	☐ VITAMINS C/D	☐ MULTIVITAMINS	☐ OTHER

WEEKLY CARE	MON	TUE	WED	THURS	FRI	SAT	SUN
REFRESH BOWL							
FRESH FOOD							
FRESH WATER							
PLAYING TIME							
EYE CHECK							
EAR CHECK							
CLAW CHECK							
TEETH CHECK							

NOTES & OBSERVATIONS

FINAL ASSESSMENT

OBJECTIVES MET	☐ YES	☐ NO	OWNER SIGNED:

RAT NAME		DATE
RAT SPECIES		
WEEK STARTING		/ /

RAT AGE		HEALTHCARE RATING	1	2	3	4	5
RAT GENDER							

TIME OF YEAR					FOOD GIVEN	

CLEAN START	
CLEAN END	
RAT SIZE	
RAT WEIGHT	

PET RAT CARE CHECKLIST

☐ CLEAN CAGE	☐ CHANGE TOYS
☐ CHANGE LINING	☐ FOOD/WATER
☐ CHANGE STRAW	☐ CLEAN FAECES

RAT VITAMINS & SUPPLEMENTS

☐ VITAMIN A	☐ VITAMINS C/D	☐ MULTIVITAMINS	☐ OTHER

WEEKLY CARE	MON	TUE	WED	THURS	FRI	SAT	SUN
REFRESH BOWL							
FRESH FOOD							
FRESH WATER							
PLAYING TIME							
EYE CHECK							
EAR CHECK							
CLAW CHECK							
TEETH CHECK							

NOTES & OBSERVATIONS

FINAL ASSESSMENT

OBJECTIVES MET ☐ YES ☐ NO OWNER SIGNED:

RAT NAME		DATE
RAT SPECIES		
WEEK STARTING		/ /

RAT AGE			HEALTHCARE RATING	1	2	3	4	5
RAT GENDER								

TIME OF YEAR					FOOD GIVEN	

CLEAN START		
CLEAN END		
RAT SIZE		
RAT WEIGHT		

PET RAT CARE CHECKLIST

☐ CLEAN CAGE	☐ CHANGE TOYS
☐ CHANGE LINING	☐ FOOD/WATER
☐ CHANGE STRAW	☐ CLEAN FAECES

RAT VITAMINS & SUPPLEMENTS

☐ VITAMIN A	☐ VITAMINS C/D	☐ MULTIVITAMINS	☐ OTHER

WEEKLY CARE	MON	TUE	WED	THURS	FRI	SAT	SUN
REFRESH BOWL							
FRESH FOOD							
FRESH WATER							
PLAYING TIME							
EYE CHECK							
EAR CHECK							
CLAW CHECK							
TEETH CHECK							

NOTES & OBSERVATIONS

FINAL ASSESSMENT

OBJECTIVES MET ☐ YES	☐ NO	OWNER SIGNED:	

RAT NAME		DATE
RAT SPECIES		
WEEK STARTING		/ /

RAT AGE		HEALTHCARE RATING	1	2	3	4	5
RAT GENDER							

TIME OF YEAR		FOOD GIVEN	
CLEAN START			
CLEAN END			
RAT SIZE			
RAT WEIGHT			

PET RAT CARE CHECKLIST

☐ CLEAN CAGE	☐ CHANGE TOYS
☐ CHANGE LINING	☐ FOOD/WATER
☐ CHANGE STRAW	☐ CLEAN FAECES

RAT VITAMINS & SUPPLEMENTS

☐ VITAMIN A	☐ VITAMINS C/D	☐ MULTIVITAMINS	☐ OTHER

WEEKLY CARE	MON	TUE	WED	THURS	FRI	SAT	SUN
REFRESH BOWL							
FRESH FOOD							
FRESH WATER							
PLAYING TIME							
EYE CHECK							
EAR CHECK							
CLAW CHECK							
TEETH CHECK							

NOTES & OBSERVATIONS

FINAL ASSESSMENT

OBJECTIVES MET ☐ YES	☐ NO	OWNER SIGNED:	

RAT NAME		DATE
RAT SPECIES		
WEEK STARTING		/ /

RAT AGE		HEALTHCARE RATING	1	2	3	4	5
RAT GENDER							
TIME OF YEAR		FOOD GIVEN					
CLEAN START							
CLEAN END							
RAT SIZE							
RAT WEIGHT							

PET RAT CARE CHECKLIST

☐ CLEAN CAGE	☐ CHANGE TOYS
☐ CHANGE LINING	☐ FOOD/WATER
☐ CHANGE STRAW	☐ CLEAN FAECES

RAT VITAMINS & SUPPLEMENTS

☐ VITAMIN A	☐ VITAMINS C/D	☐ MULTIVITAMINS	☐ OTHER

WEEKLY CARE	MON	TUE	WED	THURS	FRI	SAT	SUN
REFRESH BOWL							
FRESH FOOD							
FRESH WATER							
PLAYING TIME							
EYE CHECK							
EAR CHECK							
CLAW CHECK							
TEETH CHECK							

NOTES & OBSERVATIONS

FINAL ASSESSMENT

OBJECTIVES MET ☐ YES	☐ NO	OWNER SIGNED:	

RAT NAME		DATE
RAT SPECIES		
WEEK STARTING		/ /

RAT AGE		HEALTHCARE RATING	1	2	3	4	5
RAT GENDER							
TIME OF YEAR		FOOD GIVEN					

CLEAN START	
CLEAN END	
RAT SIZE	
RAT WEIGHT	

PET RAT CARE CHECKLIST

- ☐ CLEAN CAGE
- ☐ CHANGE TOYS
- ☐ CHANGE LINING
- ☐ FOOD/WATER
- ☐ CHANGE STRAW
- ☐ CLEAN FAECES

RAT VITAMINS & SUPPLEMENTS

☐ VITAMIN A ☐ VITAMINS C/D ☐ MULTIVITAMINS ☐ OTHER

WEEKLY CARE	MON	TUE	WED	THURS	FRI	SAT	SUN
REFRESH BOWL							
FRESH FOOD							
FRESH WATER							
PLAYING TIME							
EYE CHECK							
EAR CHECK							
CLAW CHECK							
TEETH CHECK							

NOTES & OBSERVATIONS

FINAL ASSESSMENT

OBJECTIVES MET ☐ YES ☐ NO OWNER SIGNED:

RAT NAME			DATE
RAT SPECIES			
WEEK STARTING			/ /

RAT AGE			HEALTHCARE RATING	1	2	3	4	5
RAT GENDER								

TIME OF YEAR					FOOD GIVEN	

CLEAN START		
CLEAN END		
RAT SIZE		
RAT WEIGHT		

PET RAT CARE CHECKLIST

☐ CLEAN CAGE	☐ CHANGE TOYS
☐ CHANGE LINING	☐ FOOD/WATER
☐ CHANGE STRAW	☐ CLEAN FAECES

RAT VITAMINS & SUPPLEMENTS

☐ VITAMIN A	☐ VITAMINS C/D	☐ MULTIVITAMINS	☐ OTHER

WEEKLY CARE	MON	TUE	WED	THURS	FRI	SAT	SUN
REFRESH BOWL							
FRESH FOOD							
FRESH WATER							
PLAYING TIME							
EYE CHECK							
EAR CHECK							
CLAW CHECK							
TEETH CHECK							

NOTES & OBSERVATIONS

FINAL ASSESSMENT

OBJECTIVES MET	☐ YES	☐ NO	OWNER SIGNED:

RAT NAME		DATE
RAT SPECIES		
WEEK STARTING		/ /

			HEALTHCARE RATING	1	2	3	4	5
RAT AGE								
RAT GENDER								

TIME OF YEAR						FOOD GIVEN	

CLEAN START		
CLEAN END		
RAT SIZE		
RAT WEIGHT		

PET RAT CARE CHECKLIST

☐ CLEAN CAGE	☐ CHANGE TOYS
☐ CHANGE LINING	☐ FOOD/WATER
☐ CHANGE STRAW	☐ CLEAN FAECES

RAT VITAMINS & SUPPLEMENTS

☐ VITAMIN A	☐ VITAMINS C/D	☐ MULTIVITAMINS	☐ OTHER

WEEKLY CARE	MON	TUE	WED	THURS	FRI	SAT	SUN
REFRESH BOWL							
FRESH FOOD							
FRESH WATER							
PLAYING TIME							
EYE CHECK							
EAR CHECK							
CLAW CHECK							
TEETH CHECK							

NOTES & OBSERVATIONS

FINAL ASSESSMENT

OBJECTIVES MET ☐ YES ☐ NO OWNER SIGNED:

RAT NAME		DATE
RAT SPECIES		
WEEK STARTING		/ /

RAT AGE		HEALTHCARE RATING	1	2	3	4	5
RAT GENDER							

TIME OF YEAR					FOOD GIVEN	

CLEAN START	
CLEAN END	
RAT SIZE	
RAT WEIGHT	

PET RAT CARE CHECKLIST

☐ CLEAN CAGE	☐ CHANGE TOYS
☐ CHANGE LINING	☐ FOOD/WATER
☐ CHANGE STRAW	☐ CLEAN FAECES

RAT VITAMINS & SUPPLEMENTS

☐ VITAMIN A	☐ VITAMINS C/D	☐ MULTIVITAMINS	☐ OTHER

WEEKLY CARE	MON	TUE	WED	THURS	FRI	SAT	SUN
REFRESH BOWL							
FRESH FOOD							
FRESH WATER							
PLAYING TIME							
EYE CHECK							
EAR CHECK							
CLAW CHECK							
TEETH CHECK							

NOTES & OBSERVATIONS

FINAL ASSESSMENT

OBJECTIVES MET ☐ YES	☐ NO	OWNER SIGNED:

RAT NAME		DATE
RAT SPECIES		
WEEK STARTING		/ /

RAT AGE			HEALTHCARE RATING	1	2	3	4	5
RAT GENDER								
TIME OF YEAR			FOOD GIVEN					
CLEAN START			PET RAT CARE CHECKLIST					
CLEAN END			☐ CLEAN CAGE	☐ CHANGE TOYS				
RAT SIZE			☐ CHANGE LINING	☐ FOOD/WATER				
RAT WEIGHT			☐ CHANGE STRAW	☐ CLEAN FAECES				

RAT VITAMINS & SUPPLEMENTS

☐ VITAMIN A ☐ VITAMINS C/D ☐ MULTIVITAMINS ☐ OTHER

WEEKLY CARE	MON	TUE	WED	THURS	FRI	SAT	SUN
REFRESH BOWL							
FRESH FOOD							
FRESH WATER							
PLAYING TIME							
EYE CHECK							
EAR CHECK							
CLAW CHECK							
TEETH CHECK							

NOTES & OBSERVATIONS

FINAL ASSESSMENT

OBJECTIVES MET ☐ YES ☐ NO OWNER SIGNED:

RAT NAME		DATE	
RAT SPECIES			
WEEK STARTING		/ /	

RAT AGE			HEALTHCARE RATING	1	2	3	4	5
RAT GENDER								

TIME OF YEAR		FOOD GIVEN

CLEAN START		
CLEAN END		
RAT SIZE		
RAT WEIGHT		

PET RAT CARE CHECKLIST

☐ CLEAN CAGE	☐ CHANGE TOYS
☐ CHANGE LINING	☐ FOOD/WATER
☐ CHANGE STRAW	☐ CLEAN FAECES

RAT VITAMINS & SUPPLEMENTS

☐ VITAMIN A	☐ VITAMINS C/D	☐ MULTIVITAMINS	☐ OTHER

WEEKLY CARE	MON	TUE	WED	THURS	FRI	SAT	SUN
REFRESH BOWL							
FRESH FOOD							
FRESH WATER							
PLAYING TIME							
EYE CHECK							
EAR CHECK							
CLAW CHECK							
TEETH CHECK							

NOTES & OBSERVATIONS

FINAL ASSESSMENT

OBJECTIVES MET ☐ YES ☐ NO OWNER SIGNED:

RAT NAME		DATE
RAT SPECIES		/ /
WEEK STARTING		

RAT AGE		HEALTHCARE RATING: 1 2 3 4 5
RAT GENDER		
TIME OF YEAR		FOOD GIVEN
CLEAN START		
CLEAN END		
RAT SIZE		
RAT WEIGHT		

PET RAT CARE CHECKLIST

- ☐ CLEAN CAGE
- ☐ CHANGE TOYS
- ☐ CHANGE LINING
- ☐ FOOD/WATER
- ☐ CHANGE STRAW
- ☐ CLEAN FAECES

RAT VITAMINS & SUPPLEMENTS

☐ VITAMIN A ☐ VITAMINS C/D ☐ MULTIVITAMINS ☐ OTHER

WEEKLY CARE	MON	TUE	WED	THURS	FRI	SAT	SUN
REFRESH BOWL							
FRESH FOOD							
FRESH WATER							
PLAYING TIME							
EYE CHECK							
EAR CHECK							
CLAW CHECK							
TEETH CHECK							

NOTES & OBSERVATIONS

FINAL ASSESSMENT

OBJECTIVES MET ☐ YES ☐ NO OWNER SIGNED:

RAT NAME		DATE
RAT SPECIES		
WEEK STARTING		/ /

RAT AGE		HEALTHCARE RATING	1	2	3	4	5
RAT GENDER							
TIME OF YEAR		FOOD GIVEN					
CLEAN START		PET RAT CARE CHECKLIST					
CLEAN END		☐ CLEAN CAGE	☐ CHANGE TOYS				
RAT SIZE		☐ CHANGE LINING	☐ FOOD/WATER				
RAT WEIGHT		☐ CHANGE STRAW	☐ CLEAN FAECES				

RAT VITAMINS & SUPPLEMENTS

☐ VITAMIN A ☐ VITAMINS C/D ☐ MULTIVITAMINS ☐ OTHER

WEEKLY CARE	MON	TUE	WED	THURS	FRI	SAT	SUN
REFRESH BOWL							
FRESH FOOD							
FRESH WATER							
PLAYING TIME							
EYE CHECK							
EAR CHECK							
CLAW CHECK							
TEETH CHECK							

NOTES & OBSERVATIONS

FINAL ASSESSMENT

OBJECTIVES MET ☐ YES ☐ NO OWNER SIGNED:

RAT NAME		DATE
RAT SPECIES		
WEEK STARTING		/ /

RAT AGE			HEALTHCARE RATING	1	2	3	4	5
RAT GENDER								
TIME OF YEAR			FOOD GIVEN					

CLEAN START		
CLEAN END		
RAT SIZE		
RAT WEIGHT		

PET RAT CARE CHECKLIST

☐ CLEAN CAGE	☐ CHANGE TOYS
☐ CHANGE LINING	☐ FOOD/WATER
☐ CHANGE STRAW	☐ CLEAN FAECES

RAT VITAMINS & SUPPLEMENTS

☐ VITAMIN A ☐ VITAMINS C/D ☐ MULTIVITAMINS ☐ OTHER

WEEKLY CARE	MON	TUE	WED	THURS	FRI	SAT	SUN
REFRESH BOWL							
FRESH FOOD							
FRESH WATER							
PLAYING TIME							
EYE CHECK							
EAR CHECK							
CLAW CHECK							
TEETH CHECK							

NOTES & OBSERVATIONS

FINAL ASSESSMENT

OBJECTIVES MET ☐ YES ☐ NO OWNER SIGNED:

RAT NAME		DATE
RAT SPECIES		
WEEK STARTING		/ /

RAT AGE		HEALTHCARE RATING: 1 2 3 4 5
RAT GENDER		
TIME OF YEAR		FOOD GIVEN
CLEAN START		
CLEAN END		
RAT SIZE		
RAT WEIGHT		

PET RAT CARE CHECKLIST

- ☐ CLEAN CAGE
- ☐ CHANGE TOYS
- ☐ CHANGE LINING
- ☐ FOOD/WATER
- ☐ CHANGE STRAW
- ☐ CLEAN FAECES

RAT VITAMINS & SUPPLEMENTS

☐ VITAMIN A ☐ VITAMINS C/D ☐ MULTIVITAMINS ☐ OTHER

WEEKLY CARE	MON	TUE	WED	THURS	FRI	SAT	SUN
REFRESH BOWL							
FRESH FOOD							
FRESH WATER							
PLAYING TIME							
EYE CHECK							
EAR CHECK							
CLAW CHECK							
TEETH CHECK							

NOTES & OBSERVATIONS

FINAL ASSESSMENT

OBJECTIVES MET ☐ YES ☐ NO OWNER SIGNED:

RAT NAME		DATE
RAT SPECIES		/ /
WEEK STARTING		

RAT AGE		HEALTHCARE RATING	1	2	3	4	5
RAT GENDER							

TIME OF YEAR		FOOD GIVEN

CLEAN START		PET RAT CARE CHECKLIST	
CLEAN END		☐ CLEAN CAGE	☐ CHANGE TOYS
RAT SIZE		☐ CHANGE LINING	☐ FOOD/WATER
RAT WEIGHT		☐ CHANGE STRAW	☐ CLEAN FAECES

RAT VITAMINS & SUPPLEMENTS

☐ VITAMIN A ☐ VITAMINS C/D ☐ MULTIVITAMINS ☐ OTHER

WEEKLY CARE	MON	TUE	WED	THURS	FRI	SAT	SUN
REFRESH BOWL							
FRESH FOOD							
FRESH WATER							
PLAYING TIME							
EYE CHECK							
EAR CHECK							
CLAW CHECK							
TEETH CHECK							

NOTES & OBSERVATIONS

FINAL ASSESSMENT

OBJECTIVES MET ☐ YES ☐ NO OWNER SIGNED:

RAT NAME		DATE
RAT SPECIES		
WEEK STARTING		/ /

RAT AGE			HEALTHCARE RATING	1	2	3	4	5
RAT GENDER								
TIME OF YEAR			FOOD GIVEN					
CLEAN START			PET RAT CARE CHECKLIST					
CLEAN END			☐ CLEAN CAGE		☐ CHANGE TOYS			
RAT SIZE			☐ CHANGE LINING		☐ FOOD/WATER			
RAT WEIGHT			☐ CHANGE STRAW		☐ CLEAN FAECES			

RAT VITAMINS & SUPPLEMENTS

☐ VITAMIN A ☐ VITAMINS C/D ☐ MULTIVITAMINS ☐ OTHER

WEEKLY CARE	MON	TUE	WED	THURS	FRI	SAT	SUN
REFRESH BOWL							
FRESH FOOD							
FRESH WATER							
PLAYING TIME							
EYE CHECK							
EAR CHECK							
CLAW CHECK							
TEETH CHECK							

NOTES & OBSERVATIONS

FINAL ASSESSMENT

OBJECTIVES MET ☐ YES ☐ NO OWNER SIGNED:

RAT NAME		DATE
RAT SPECIES		
WEEK STARTING		/ /

RAT AGE			HEALTHCARE RATING	1	2	3	4	5
RAT GENDER								
TIME OF YEAR			FOOD GIVEN					
CLEAN START								
CLEAN END								
RAT SIZE								
RAT WEIGHT								

PET RAT CARE CHECKLIST

☐ CLEAN CAGE	☐ CHANGE TOYS
☐ CHANGE LINING	☐ FOOD/WATER
☐ CHANGE STRAW	☐ CLEAN FAECES

RAT VITAMINS & SUPPLEMENTS

☐ VITAMIN A ☐ VITAMINS C/D ☐ MULTIVITAMINS ☐ OTHER

WEEKLY CARE	MON	TUE	WED	THURS	FRI	SAT	SUN
REFRESH BOWL							
FRESH FOOD							
FRESH WATER							
PLAYING TIME							
EYE CHECK							
EAR CHECK							
CLAW CHECK							
TEETH CHECK							

NOTES & OBSERVATIONS

FINAL ASSESSMENT

OBJECTIVES MET ☐ YES ☐ NO OWNER SIGNED:

RAT NAME		DATE
RAT SPECIES		
WEEK STARTING		/ /

RAT AGE			HEALTHCARE RATING	1	2	3	4	5
RAT GENDER								
TIME OF YEAR			FOOD GIVEN					
CLEAN START								
CLEAN END								
RAT SIZE								
RAT WEIGHT								

PET RAT CARE CHECKLIST

- ☐ CLEAN CAGE
- ☐ CHANGE TOYS
- ☐ CHANGE LINING
- ☐ FOOD/WATER
- ☐ CHANGE STRAW
- ☐ CLEAN FAECES

RAT VITAMINS & SUPPLEMENTS

☐ VITAMIN A ☐ VITAMINS C/D ☐ MULTIVITAMINS ☐ OTHER

WEEKLY CARE	MON	TUE	WED	THURS	FRI	SAT	SUN
REFRESH BOWL							
FRESH FOOD							
FRESH WATER							
PLAYING TIME							
EYE CHECK							
EAR CHECK							
CLAW CHECK							
TEETH CHECK							

NOTES & OBSERVATIONS

FINAL ASSESSMENT

OBJECTIVES MET ☐ YES ☐ NO OWNER SIGNED:

RAT NAME	
RAT SPECIES	
WEEK STARTING	

DATE

/ /

RAT AGE		HEALTHCARE RATING	1	2	3	4	5
RAT GENDER							

TIME OF YEAR

FOOD GIVEN

CLEAN START	
CLEAN END	
RAT SIZE	
RAT WEIGHT	

PET RAT CARE CHECKLIST

- ☐ CLEAN CAGE
- ☐ CHANGE TOYS
- ☐ CHANGE LINING
- ☐ FOOD/WATER
- ☐ CHANGE STRAW
- ☐ CLEAN FAECES

RAT VITAMINS & SUPPLEMENTS

- ☐ VITAMIN A
- ☐ VITAMINS C/D
- ☐ MULTIVITAMINS
- ☐ OTHER

WEEKLY CARE	MON	TUE	WED	THURS	FRI	SAT	SUN
REFRESH BOWL							
FRESH FOOD							
FRESH WATER							
PLAYING TIME							
EYE CHECK							
EAR CHECK							
CLAW CHECK							
TEETH CHECK							

NOTES & OBSERVATIONS

FINAL ASSESSMENT

OBJECTIVES MET ☐ YES ☐ NO OWNER SIGNED:

RAT NAME		DATE	
RAT SPECIES			
WEEK STARTING		/ /	

RAT AGE			HEALTHCARE RATING	1	2	3	4	5
RAT GENDER								

TIME OF YEAR						FOOD GIVEN	

CLEAN START		PET RAT CARE CHECKLIST	
CLEAN END		☐ CLEAN CAGE	☐ CHANGE TOYS
RAT SIZE		☐ CHANGE LINING	☐ FOOD/WATER
RAT WEIGHT		☐ CHANGE STRAW	☐ CLEAN FAECES

RAT VITAMINS & SUPPLEMENTS

☐ VITAMIN A ☐ VITAMINS C/D ☐ MULTIVITAMINS ☐ OTHER

WEEKLY CARE	MON	TUE	WED	THURS	FRI	SAT	SUN
REFRESH BOWL							
FRESH FOOD							
FRESH WATER							
PLAYING TIME							
EYE CHECK							
EAR CHECK							
CLAW CHECK							
TEETH CHECK							

NOTES & OBSERVATIONS

FINAL ASSESSMENT

OBJECTIVES MET ☐ YES ☐ NO OWNER SIGNED:

RAT NAME		DATE
RAT SPECIES		
WEEK STARTING		/ /

RAT AGE			HEALTHCARE RATING	1	2	3	4	5
RAT GENDER								

TIME OF YEAR						FOOD GIVEN	

CLEAN START			
CLEAN END			
RAT SIZE			
RAT WEIGHT			

PET RAT CARE CHECKLIST

☐ CLEAN CAGE	☐ CHANGE TOYS
☐ CHANGE LINING	☐ FOOD/WATER
☐ CHANGE STRAW	☐ CLEAN FAECES

RAT VITAMINS & SUPPLEMENTS

☐ VITAMIN A	☐ VITAMINS C/D	☐ MULTIVITAMINS	☐ OTHER

WEEKLY CARE	MON	TUE	WED	THURS	FRI	SAT	SUN
REFRESH BOWL							
FRESH FOOD							
FRESH WATER							
PLAYING TIME							
EYE CHECK							
EAR CHECK							
CLAW CHECK							
TEETH CHECK							

NOTES & OBSERVATIONS

FINAL ASSESSMENT

OBJECTIVES MET	☐ YES	☐ NO	OWNER SIGNED:

RAT NAME		DATE
RAT SPECIES		
WEEK STARTING		/ /

RAT AGE		HEALTHCARE RATING	1	2	3	4	5
RAT GENDER							

TIME OF YEAR	
FOOD GIVEN	

CLEAN START	
CLEAN END	
RAT SIZE	
RAT WEIGHT	

PET RAT CARE CHECKLIST

- ☐ CLEAN CAGE
- ☐ CHANGE TOYS
- ☐ CHANGE LINING
- ☐ FOOD/WATER
- ☐ CHANGE STRAW
- ☐ CLEAN FAECES

RAT VITAMINS & SUPPLEMENTS

☐ VITAMIN A ☐ VITAMINS C/D ☐ MULTIVITAMINS ☐ OTHER

WEEKLY CARE	MON	TUE	WED	THURS	FRI	SAT	SUN
REFRESH BOWL							
FRESH FOOD							
FRESH WATER							
PLAYING TIME							
EYE CHECK							
EAR CHECK							
CLAW CHECK							
TEETH CHECK							

NOTES & OBSERVATIONS

FINAL ASSESSMENT

OBJECTIVES MET ☐ YES ☐ NO OWNER SIGNED:

RAT NAME		DATE
RAT SPECIES		
WEEK STARTING		/ /

RAT AGE		HEALTHCARE RATING	1	2	3	4	5
RAT GENDER							

TIME OF YEAR		FOOD GIVEN	
CLEAN START		PET RAT CARE CHECKLIST	
CLEAN END		☐ CLEAN CAGE	☐ CHANGE TOYS
RAT SIZE		☐ CHANGE LINING	☐ FOOD/WATER
RAT WEIGHT		☐ CHANGE STRAW	☐ CLEAN FAECES

RAT VITAMINS & SUPPLEMENTS

☐ VITAMIN A ☐ VITAMINS C/D ☐ MULTIVITAMINS ☐ OTHER

WEEKLY CARE	MON	TUE	WED	THURS	FRI	SAT	SUN
REFRESH BOWL							
FRESH FOOD							
FRESH WATER							
PLAYING TIME							
EYE CHECK							
EAR CHECK							
CLAW CHECK							
TEETH CHECK							

NOTES & OBSERVATIONS

FINAL ASSESSMENT

OBJECTIVES MET ☐ YES ☐ NO OWNER SIGNED:

RAT NAME		DATE
RAT SPECIES		
WEEK STARTING		/ /

RAT AGE		HEALTHCARE RATING	1	2	3	4	5
RAT GENDER							
TIME OF YEAR		FOOD GIVEN					
CLEAN START		PET RAT CARE CHECKLIST					
CLEAN END		☐ CLEAN CAGE	☐ CHANGE TOYS				
RAT SIZE		☐ CHANGE LINING	☐ FOOD/WATER				
RAT WEIGHT		☐ CHANGE STRAW	☐ CLEAN FAECES				

RAT VITAMINS & SUPPLEMENTS

☐ VITAMIN A ☐ VITAMINS C/D ☐ MULTIVITAMINS ☐ OTHER

WEEKLY CARE	MON	TUE	WED	THURS	FRI	SAT	SUN
REFRESH BOWL							
FRESH FOOD							
FRESH WATER							
PLAYING TIME							
EYE CHECK							
EAR CHECK							
CLAW CHECK							
TEETH CHECK							

NOTES & OBSERVATIONS

FINAL ASSESSMENT

OBJECTIVES MET ☐ YES ☐ NO OWNER SIGNED:

RAT NAME		DATE
RAT SPECIES		
WEEK STARTING		/ /

RAT AGE		HEALTHCARE RATING	1	2	3	4	5
RAT GENDER							

TIME OF YEAR					FOOD GIVEN	

CLEAN START		PET RAT CARE CHECKLIST	
CLEAN END		☐ CLEAN CAGE	☐ CHANGE TOYS
RAT SIZE		☐ CHANGE LINING	☐ FOOD/WATER
RAT WEIGHT		☐ CHANGE STRAW	☐ CLEAN FAECES

RAT VITAMINS & SUPPLEMENTS

☐ VITAMIN A ☐ VITAMINS C/D ☐ MULTIVITAMINS ☐ OTHER

WEEKLY CARE	MON	TUE	WED	THURS	FRI	SAT	SUN
REFRESH BOWL							
FRESH FOOD							
FRESH WATER							
PLAYING TIME							
EYE CHECK							
EAR CHECK							
CLAW CHECK							
TEETH CHECK							

NOTES & OBSERVATIONS

FINAL ASSESSMENT

OBJECTIVES MET ☐ YES ☐ NO OWNER SIGNED:

RAT NAME		DATE
RAT SPECIES		/ /
WEEK STARTING		

RAT AGE	🐀		HEALTHCARE RATING	1 2 3 4 5
RAT GENDER				
TIME OF YEAR	🌷 ☀ 🍂 ❄		FOOD GIVEN	
CLEAN START	🕒		**PET RAT CARE CHECKLIST**	
CLEAN END			☐ CLEAN CAGE	☐ CHANGE TOYS
RAT SIZE	⬍		☐ CHANGE LINING	☐ FOOD/WATER
RAT WEIGHT			☐ CHANGE STRAW	☐ CLEAN FAECES

RAT VITAMINS & SUPPLEMENTS

☐ VITAMIN A ☐ VITAMINS C/D ☐ MULTIVITAMINS ☐ OTHER

WEEKLY CARE	MON	TUE	WED	THURS	FRI	SAT	SUN
REFRESH BOWL							
FRESH FOOD							
FRESH WATER							
PLAYING TIME							
EYE CHECK							
EAR CHECK							
CLAW CHECK							
TEETH CHECK							

NOTES & OBSERVATIONS

FINAL ASSESSMENT

OBJECTIVES MET ☐ YES ☐ NO OWNER SIGNED:

RAT NAME		DATE
RAT SPECIES		
WEEK STARTING		/ /

RAT AGE			HEALTHCARE RATING	1	2	3	4	5
RAT GENDER								
TIME OF YEAR					FOOD GIVEN			
CLEAN START								
CLEAN END								
RAT SIZE								
RAT WEIGHT								

PET RAT CARE CHECKLIST

☐ CLEAN CAGE	☐ CHANGE TOYS
☐ CHANGE LINING	☐ FOOD/WATER
☐ CHANGE STRAW	☐ CLEAN FAECES

RAT VITAMINS & SUPPLEMENTS

☐ VITAMIN A	☐ VITAMINS C/D	☐ MULTIVITAMINS	☐ OTHER

WEEKLY CARE	MON	TUE	WED	THURS	FRI	SAT	SUN
REFRESH BOWL							
FRESH FOOD							
FRESH WATER							
PLAYING TIME							
EYE CHECK							
EAR CHECK							
CLAW CHECK							
TEETH CHECK							

NOTES & OBSERVATIONS

FINAL ASSESSMENT

OBJECTIVES MET	☐ YES	☐ NO	OWNER SIGNED:

RAT NAME		DATE
RAT SPECIES		
WEEK STARTING		/ /

RAT AGE		HEALTHCARE RATING	1	2	3	4	5
RAT GENDER							

TIME OF YEAR		FOOD GIVEN	

CLEAN START		PET RAT CARE CHECKLIST	
CLEAN END		☐ CLEAN CAGE	☐ CHANGE TOYS
RAT SIZE		☐ CHANGE LINING	☐ FOOD/WATER
RAT WEIGHT		☐ CHANGE STRAW	☐ CLEAN FAECES

RAT VITAMINS & SUPPLEMENTS

☐ VITAMIN A ☐ VITAMINS C/D ☐ MULTIVITAMINS ☐ OTHER

WEEKLY CARE	MON	TUE	WED	THURS	FRI	SAT	SUN
REFRESH BOWL							
FRESH FOOD							
FRESH WATER							
PLAYING TIME							
EYE CHECK							
EAR CHECK							
CLAW CHECK							
TEETH CHECK							

NOTES & OBSERVATIONS

FINAL ASSESSMENT

OBJECTIVES MET ☐ YES ☐ NO OWNER SIGNED:

RAT NAME		DATE	
RAT SPECIES		/ /	
WEEK STARTING			

RAT AGE		HEALTHCARE RATING	1	2	3	4	5
RAT GENDER							
TIME OF YEAR		FOOD GIVEN					
CLEAN START							
CLEAN END							
RAT SIZE							
RAT WEIGHT							

PET RAT CARE CHECKLIST

- ☐ CLEAN CAGE
- ☐ CHANGE TOYS
- ☐ CHANGE LINING
- ☐ FOOD/WATER
- ☐ CHANGE STRAW
- ☐ CLEAN FAECES

RAT VITAMINS & SUPPLEMENTS

☐ VITAMIN A ☐ VITAMINS C/D ☐ MULTIVITAMINS ☐ OTHER

WEEKLY CARE	MON	TUE	WED	THURS	FRI	SAT	SUN
REFRESH BOWL							
FRESH FOOD							
FRESH WATER							
PLAYING TIME							
EYE CHECK							
EAR CHECK							
CLAW CHECK							
TEETH CHECK							

NOTES & OBSERVATIONS

FINAL ASSESSMENT

OBJECTIVES MET ☐ YES ☐ NO OWNER SIGNED:

RAT NAME		DATE
RAT SPECIES		
WEEK STARTING		/ /

RAT AGE		HEALTHCARE RATING	1	2	3	4	5
RAT GENDER							

TIME OF YEAR

FOOD GIVEN

CLEAN START	
CLEAN END	
RAT SIZE	
RAT WEIGHT	

PET RAT CARE CHECKLIST

- ☐ CLEAN CAGE
- ☐ CHANGE TOYS
- ☐ CHANGE LINING
- ☐ FOOD/WATER
- ☐ CHANGE STRAW
- ☐ CLEAN FAECES

RAT VITAMINS & SUPPLEMENTS

☐ VITAMIN A ☐ VITAMINS C/D ☐ MULTIVITAMINS ☐ OTHER

WEEKLY CARE	MON	TUE	WED	THURS	FRI	SAT	SUN
REFRESH BOWL							
FRESH FOOD							
FRESH WATER							
PLAYING TIME							
EYE CHECK							
EAR CHECK							
CLAW CHECK							
TEETH CHECK							

NOTES & OBSERVATIONS

FINAL ASSESSMENT

OBJECTIVES MET ☐ YES ☐ NO OWNER SIGNED:

RAT NAME		DATE
RAT SPECIES		/ /
WEEK STARTING		

RAT AGE		HEALTHCARE RATING 1 2 3 4 5
RAT GENDER		
TIME OF YEAR		FOOD GIVEN
CLEAN START		
CLEAN END		
RAT SIZE		
RAT WEIGHT		

PET RAT CARE CHECKLIST

☐ CLEAN CAGE	☐ CHANGE TOYS
☐ CHANGE LINING	☐ FOOD/WATER
☐ CHANGE STRAW	☐ CLEAN FAECES

RAT VITAMINS & SUPPLEMENTS

☐ VITAMIN A ☐ VITAMINS C/D ☐ MULTIVITAMINS ☐ OTHER

WEEKLY CARE	MON	TUE	WED	THURS	FRI	SAT	SUN
REFRESH BOWL							
FRESH FOOD							
FRESH WATER							
PLAYING TIME							
EYE CHECK							
EAR CHECK							
CLAW CHECK							
TEETH CHECK							

NOTES & OBSERVATIONS

FINAL ASSESSMENT

OBJECTIVES MET ☐ YES ☐ NO OWNER SIGNED:

RAT NAME		DATE
RAT SPECIES		
WEEK STARTING		/ /

RAT AGE		HEALTHCARE RATING	1	2	3	4	5
RAT GENDER							

TIME OF YEAR					FOOD GIVEN	

CLEAN START		PET RAT CARE CHECKLIST	
CLEAN END		☐ CLEAN CAGE	☐ CHANGE TOYS
RAT SIZE		☐ CHANGE LINING	☐ FOOD/WATER
RAT WEIGHT		☐ CHANGE STRAW	☐ CLEAN FAECES

RAT VITAMINS & SUPPLEMENTS

☐ VITAMIN A ☐ VITAMINS C/D ☐ MULTIVITAMINS ☐ OTHER

WEEKLY CARE	MON	TUE	WED	THURS	FRI	SAT	SUN
REFRESH BOWL							
FRESH FOOD							
FRESH WATER							
PLAYING TIME							
EYE CHECK							
EAR CHECK							
CLAW CHECK							
TEETH CHECK							

NOTES & OBSERVATIONS

FINAL ASSESSMENT

OBJECTIVES MET ☐ YES ☐ NO OWNER SIGNED:

RAT NAME		DATE
RAT SPECIES		
WEEK STARTING		/ /

RAT AGE		HEALTHCARE RATING	1	2	3	4	5
RAT GENDER							

TIME OF YEAR					FOOD GIVEN

CLEAN START		
CLEAN END		
RAT SIZE		
RAT WEIGHT		

PET RAT CARE CHECKLIST

☐ CLEAN CAGE	☐ CHANGE TOYS
☐ CHANGE LINING	☐ FOOD/WATER
☐ CHANGE STRAW	☐ CLEAN FAECES

RAT VITAMINS & SUPPLEMENTS

☐ VITAMIN A ☐ VITAMINS C/D ☐ MULTIVITAMINS ☐ OTHER

WEEKLY CARE	MON	TUE	WED	THURS	FRI	SAT	SUN
REFRESH BOWL							
FRESH FOOD							
FRESH WATER							
PLAYING TIME							
EYE CHECK							
EAR CHECK							
CLAW CHECK							
TEETH CHECK							

NOTES & OBSERVATIONS

FINAL ASSESSMENT

OBJECTIVES MET ☐ YES ☐ NO OWNER SIGNED:

RAT NAME		DATE
RAT SPECIES		
WEEK STARTING		/ /

RAT AGE			HEALTHCARE RATING	1	2	3	4	5
RAT GENDER								

TIME OF YEAR					FOOD GIVEN

CLEAN START		
CLEAN END		
RAT SIZE		
RAT WEIGHT		

PET RAT CARE CHECKLIST

☐ CLEAN CAGE	☐ CHANGE TOYS
☐ CHANGE LINING	☐ FOOD/WATER
☐ CHANGE STRAW	☐ CLEAN FAECES

RAT VITAMINS & SUPPLEMENTS

☐ VITAMIN A	☐ VITAMINS C/D	☐ MULTIVITAMINS	☐ OTHER

WEEKLY CARE	MON	TUE	WED	THURS	FRI	SAT	SUN
REFRESH BOWL							
FRESH FOOD							
FRESH WATER							
PLAYING TIME							
EYE CHECK							
EAR CHECK							
CLAW CHECK							
TEETH CHECK							

NOTES & OBSERVATIONS

FINAL ASSESSMENT

OBJECTIVES MET ☐ YES	☐ NO	OWNER SIGNED:	

RAT NAME		DATE
RAT SPECIES		
WEEK STARTING		/ /

RAT AGE			HEALTHCARE RATING	1	2	3	4	5
RAT GENDER								

TIME OF YEAR					FOOD GIVEN	

CLEAN START		
CLEAN END		
RAT SIZE		
RAT WEIGHT		

PET RAT CARE CHECKLIST

☐ CLEAN CAGE	☐ CHANGE TOYS
☐ CHANGE LINING	☐ FOOD/WATER
☐ CHANGE STRAW	☐ CLEAN FAECES

RAT VITAMINS & SUPPLEMENTS

☐ VITAMIN A	☐ VITAMINS C/D	☐ MULTIVITAMINS	☐ OTHER

WEEKLY CARE	MON	TUE	WED	THURS	FRI	SAT	SUN
REFRESH BOWL							
FRESH FOOD							
FRESH WATER							
PLAYING TIME							
EYE CHECK							
EAR CHECK							
CLAW CHECK							
TEETH CHECK							

NOTES & OBSERVATIONS

FINAL ASSESSMENT

OBJECTIVES MET ☐ YES	☐ NO	OWNER SIGNED:	

RAT NAME		DATE
RAT SPECIES		
WEEK STARTING		/ /

RAT AGE		HEALTHCARE RATING	1	2	3	4	5
RAT GENDER							

TIME OF YEAR					FOOD GIVEN	

CLEAN START		PET RAT CARE CHECKLIST	
CLEAN END		☐ CLEAN CAGE	☐ CHANGE TOYS
RAT SIZE		☐ CHANGE LINING	☐ FOOD/WATER
RAT WEIGHT		☐ CHANGE STRAW	☐ CLEAN FAECES

RAT VITAMINS & SUPPLEMENTS

☐ VITAMIN A ☐ VITAMINS C/D ☐ MULTIVITAMINS ☐ OTHER

WEEKLY CARE	MON	TUE	WED	THURS	FRI	SAT	SUN
REFRESH BOWL							
FRESH FOOD							
FRESH WATER							
PLAYING TIME							
EYE CHECK							
EAR CHECK							
CLAW CHECK							
TEETH CHECK							

NOTES & OBSERVATIONS

FINAL ASSESSMENT

OBJECTIVES MET ☐ YES ☐ NO OWNER SIGNED:

RAT NAME		DATE
RAT SPECIES		
WEEK STARTING		/ /

RAT AGE		HEALTHCARE RATING	1	2	3	4	5
RAT GENDER							

TIME OF YEAR		FOOD GIVEN
CLEAN START		
CLEAN END		
RAT SIZE		
RAT WEIGHT		

PET RAT CARE CHECKLIST	
☐ CLEAN CAGE	☐ CHANGE TOYS
☐ CHANGE LINING	☐ FOOD/WATER
☐ CHANGE STRAW	☐ CLEAN FAECES

RAT VITAMINS & SUPPLEMENTS

☐ VITAMIN A ☐ VITAMINS C/D ☐ MULTIVITAMINS ☐ OTHER

WEEKLY CARE	MON	TUE	WED	THURS	FRI	SAT	SUN
REFRESH BOWL							
FRESH FOOD							
FRESH WATER							
PLAYING TIME							
EYE CHECK							
EAR CHECK							
CLAW CHECK							
TEETH CHECK							

NOTES & OBSERVATIONS

FINAL ASSESSMENT

OBJECTIVES MET ☐ YES ☐ NO OWNER SIGNED:

RAT NAME		DATE	
RAT SPECIES			
WEEK STARTING		/ /	

RAT AGE		HEALTHCARE RATING	1	2	3	4	5
RAT GENDER							

TIME OF YEAR					FOOD GIVEN

CLEAN START		PET RAT CARE CHECKLIST	
CLEAN END		☐ CLEAN CAGE	☐ CHANGE TOYS
RAT SIZE		☐ CHANGE LINING	☐ FOOD/WATER
RAT WEIGHT		☐ CHANGE STRAW	☐ CLEAN FAECES

RAT VITAMINS & SUPPLEMENTS

☐ VITAMIN A ☐ VITAMINS C/D ☐ MULTIVITAMINS ☐ OTHER

WEEKLY CARE	MON	TUE	WED	THURS	FRI	SAT	SUN
REFRESH BOWL							
FRESH FOOD							
FRESH WATER							
PLAYING TIME							
EYE CHECK							
EAR CHECK							
CLAW CHECK							
TEETH CHECK							

NOTES & OBSERVATIONS

FINAL ASSESSMENT

OBJECTIVES MET ☐ YES ☐ NO OWNER SIGNED:

RAT NAME		DATE
RAT SPECIES		
WEEK STARTING		/ /

RAT AGE		HEALTHCARE RATING 1 2 3 4 5
RAT GENDER		
TIME OF YEAR		FOOD GIVEN
CLEAN START		
CLEAN END		
RAT SIZE		
RAT WEIGHT		

PET RAT CARE CHECKLIST

- ☐ CLEAN CAGE
- ☐ CHANGE TOYS
- ☐ CHANGE LINING
- ☐ FOOD/WATER
- ☐ CHANGE STRAW
- ☐ CLEAN FAECES

RAT VITAMINS & SUPPLEMENTS

☐ VITAMIN A ☐ VITAMINS C/D ☐ MULTIVITAMINS ☐ OTHER

WEEKLY CARE	MON	TUE	WED	THURS	FRI	SAT	SUN
REFRESH BOWL							
FRESH FOOD							
FRESH WATER							
PLAYING TIME							
EYE CHECK							
EAR CHECK							
CLAW CHECK							
TEETH CHECK							

NOTES & OBSERVATIONS

FINAL ASSESSMENT

OBJECTIVES MET ☐ YES ☐ NO OWNER SIGNED:

RAT NAME		DATE
RAT SPECIES		
WEEK STARTING		/ /

RAT AGE			HEALTHCARE RATING	1	2	3	4	5
RAT GENDER								

TIME OF YEAR		FOOD GIVEN
CLEAN START		
CLEAN END		
RAT SIZE		
RAT WEIGHT		

PET RAT CARE CHECKLIST

☐ CLEAN CAGE	☐ CHANGE TOYS
☐ CHANGE LINING	☐ FOOD/WATER
☐ CHANGE STRAW	☐ CLEAN FAECES

RAT VITAMINS & SUPPLEMENTS

☐ VITAMIN A ☐ VITAMINS C/D ☐ MULTIVITAMINS ☐ OTHER

WEEKLY CARE	MON	TUE	WED	THURS	FRI	SAT	SUN
REFRESH BOWL							
FRESH FOOD							
FRESH WATER							
PLAYING TIME							
EYE CHECK							
EAR CHECK							
CLAW CHECK							
TEETH CHECK							

NOTES & OBSERVATIONS

FINAL ASSESSMENT

OBJECTIVES MET ☐ YES ☐ NO OWNER SIGNED:

RAT NAME		DATE	
RAT SPECIES			
WEEK STARTING		/ /	

RAT AGE		HEALTHCARE RATING	1	2	3	4	5
RAT GENDER							
TIME OF YEAR		FOOD GIVEN					
CLEAN START							
CLEAN END							
RAT SIZE							
RAT WEIGHT							

PET RAT CARE CHECKLIST

- ☐ CLEAN CAGE
- ☐ CHANGE TOYS
- ☐ CHANGE LINING
- ☐ FOOD/WATER
- ☐ CHANGE STRAW
- ☐ CLEAN FAECES

RAT VITAMINS & SUPPLEMENTS

☐ VITAMIN A ☐ VITAMINS C/D ☐ MULTIVITAMINS ☐ OTHER

WEEKLY CARE	MON	TUE	WED	THURS	FRI	SAT	SUN
REFRESH BOWL							
FRESH FOOD							
FRESH WATER							
PLAYING TIME							
EYE CHECK							
EAR CHECK							
CLAW CHECK							
TEETH CHECK							

NOTES & OBSERVATIONS

FINAL ASSESSMENT

OBJECTIVES MET ☐ YES ☐ NO OWNER SIGNED:

RAT NAME		DATE
RAT SPECIES		
WEEK STARTING		/ /

RAT AGE			HEALTHCARE RATING	1	2	3	4	5
RAT GENDER								

TIME OF YEAR		FOOD GIVEN
CLEAN START		
CLEAN END		
RAT SIZE		
RAT WEIGHT		

PET RAT CARE CHECKLIST

☐ CLEAN CAGE	☐ CHANGE TOYS
☐ CHANGE LINING	☐ FOOD/WATER
☐ CHANGE STRAW	☐ CLEAN FAECES

RAT VITAMINS & SUPPLEMENTS

☐ VITAMIN A	☐ VITAMINS C/D	☐ MULTIVITAMINS	☐ OTHER

WEEKLY CARE	MON	TUE	WED	THURS	FRI	SAT	SUN
REFRESH BOWL							
FRESH FOOD							
FRESH WATER							
PLAYING TIME							
EYE CHECK							
EAR CHECK							
CLAW CHECK							
TEETH CHECK							

NOTES & OBSERVATIONS

FINAL ASSESSMENT

OBJECTIVES MET ☐ YES	☐ NO	OWNER SIGNED:	

RAT NAME		DATE
RAT SPECIES		
WEEK STARTING		/ /

RAT AGE			HEALTHCARE RATING	1	2	3	4	5
RAT GENDER								

TIME OF YEAR						FOOD GIVEN	

CLEAN START			PET RAT CARE CHECKLIST	
CLEAN END			☐ CLEAN CAGE	☐ CHANGE TOYS
RAT SIZE			☐ CHANGE LINING	☐ FOOD/WATER
RAT WEIGHT			☐ CHANGE STRAW	☐ CLEAN FAECES

RAT VITAMINS & SUPPLEMENTS

☐ VITAMIN A	☐ VITAMINS C/D	☐ MULTIVITAMINS	☐ OTHER

WEEKLY CARE	MON	TUE	WED	THURS	FRI	SAT	SUN
REFRESH BOWL							
FRESH FOOD							
FRESH WATER							
PLAYING TIME							
EYE CHECK							
EAR CHECK							
CLAW CHECK							
TEETH CHECK							

NOTES & OBSERVATIONS

FINAL ASSESSMENT

OBJECTIVES MET ☐ YES	☐ NO	OWNER SIGNED:	

RAT NAME		DATE
RAT SPECIES		
WEEK STARTING		/ /

			HEALTHCARE RATING	1	2	3	4	5
RAT AGE								
RAT GENDER								

TIME OF YEAR					FOOD GIVEN	

CLEAN START		
CLEAN END		
RAT SIZE		
RAT WEIGHT		

PET RAT CARE CHECKLIST

☐ CLEAN CAGE	☐ CHANGE TOYS
☐ CHANGE LINING	☐ FOOD/WATER
☐ CHANGE STRAW	☐ CLEAN FAECES

RAT VITAMINS & SUPPLEMENTS

☐ VITAMIN A ☐ VITAMINS C/D ☐ MULTIVITAMINS ☐ OTHER

WEEKLY CARE	MON	TUE	WED	THURS	FRI	SAT	SUN
REFRESH BOWL							
FRESH FOOD							
FRESH WATER							
PLAYING TIME							
EYE CHECK							
EAR CHECK							
CLAW CHECK							
TEETH CHECK							

NOTES & OBSERVATIONS

FINAL ASSESSMENT

OBJECTIVES MET ☐ YES ☐ NO OWNER SIGNED:

RAT NAME		DATE
RAT SPECIES		
WEEK STARTING		/ /

RAT AGE		HEALTHCARE RATING	1	2	3	4	5
RAT GENDER							

TIME OF YEAR					FOOD GIVEN	

CLEAN START		
CLEAN END		
RAT SIZE		
RAT WEIGHT		

PET RAT CARE CHECKLIST

☐ CLEAN CAGE	☐ CHANGE TOYS
☐ CHANGE LINING	☐ FOOD/WATER
☐ CHANGE STRAW	☐ CLEAN FAECES

RAT VITAMINS & SUPPLEMENTS

☐ VITAMIN A ☐ VITAMINS C/D ☐ MULTIVITAMINS ☐ OTHER

WEEKLY CARE	MON	TUE	WED	THURS	FRI	SAT	SUN
REFRESH BOWL							
FRESH FOOD							
FRESH WATER							
PLAYING TIME							
EYE CHECK							
EAR CHECK							
CLAW CHECK							
TEETH CHECK							

NOTES & OBSERVATIONS

FINAL ASSESSMENT

OBJECTIVES MET ☐ YES ☐ NO OWNER SIGNED:

RAT NAME		DATE
RAT SPECIES		
WEEK STARTING		/ /

RAT AGE			HEALTHCARE RATING	1	2	3	4	5
RAT GENDER								
TIME OF YEAR			FOOD GIVEN					

CLEAN START		
CLEAN END		
RAT SIZE		
RAT WEIGHT		

PET RAT CARE CHECKLIST

☐ CLEAN CAGE	☐ CHANGE TOYS
☐ CHANGE LINING	☐ FOOD/WATER
☐ CHANGE STRAW	☐ CLEAN FAECES

RAT VITAMINS & SUPPLEMENTS

☐ VITAMIN A	☐ VITAMINS C/D	☐ MULTIVITAMINS	☐ OTHER

WEEKLY CARE	MON	TUE	WED	THURS	FRI	SAT	SUN
REFRESH BOWL							
FRESH FOOD							
FRESH WATER							
PLAYING TIME							
EYE CHECK							
EAR CHECK							
CLAW CHECK							
TEETH CHECK							

NOTES & OBSERVATIONS

FINAL ASSESSMENT

OBJECTIVES MET ☐ YES	☐ NO	OWNER SIGNED:	

RAT NAME		DATE
RAT SPECIES		
WEEK STARTING		/ /

RAT AGE		HEALTHCARE RATING	1	2	3	4	5
RAT GENDER							

TIME OF YEAR					FOOD GIVEN	

CLEAN START	
CLEAN END	
RAT SIZE	
RAT WEIGHT	

PET RAT CARE CHECKLIST

- ☐ CLEAN CAGE
- ☐ CHANGE TOYS
- ☐ CHANGE LINING
- ☐ FOOD/WATER
- ☐ CHANGE STRAW
- ☐ CLEAN FAECES

RAT VITAMINS & SUPPLEMENTS

☐ VITAMIN A ☐ VITAMINS C/D ☐ MULTIVITAMINS ☐ OTHER

WEEKLY CARE	MON	TUE	WED	THURS	FRI	SAT	SUN
REFRESH BOWL							
FRESH FOOD							
FRESH WATER							
PLAYING TIME							
EYE CHECK							
EAR CHECK							
CLAW CHECK							
TEETH CHECK							

NOTES & OBSERVATIONS

FINAL ASSESSMENT

OBJECTIVES MET ☐ YES ☐ NO OWNER SIGNED:

RAT NAME		DATE
RAT SPECIES		
WEEK STARTING		/ /

RAT AGE		HEALTHCARE RATING	1	2	3	4	5
RAT GENDER							

TIME OF YEAR	
FOOD GIVEN	

CLEAN START	
CLEAN END	
RAT SIZE	
RAT WEIGHT	

PET RAT CARE CHECKLIST

☐ CLEAN CAGE	☐ CHANGE TOYS
☐ CHANGE LINING	☐ FOOD/WATER
☐ CHANGE STRAW	☐ CLEAN FAECES

RAT VITAMINS & SUPPLEMENTS

☐ VITAMIN A	☐ VITAMINS C/D	☐ MULTIVITAMINS	☐ OTHER

WEEKLY CARE	MON	TUE	WED	THURS	FRI	SAT	SUN
REFRESH BOWL							
FRESH FOOD							
FRESH WATER							
PLAYING TIME							
EYE CHECK							
EAR CHECK							
CLAW CHECK							
TEETH CHECK							

NOTES & OBSERVATIONS

FINAL ASSESSMENT

OBJECTIVES MET	☐ YES	☐ NO	OWNER SIGNED:

RAT NAME		DATE
RAT SPECIES		
WEEK STARTING		/ /

RAT AGE		HEALTHCARE RATING	1	2	3	4	5
RAT GENDER							

TIME OF YEAR		FOOD GIVEN

CLEAN START		PET RAT CARE CHECKLIST	
CLEAN END		☐ CLEAN CAGE	☐ CHANGE TOYS
RAT SIZE		☐ CHANGE LINING	☐ FOOD/WATER
RAT WEIGHT		☐ CHANGE STRAW	☐ CLEAN FAECES

RAT VITAMINS & SUPPLEMENTS

☐ VITAMIN A	☐ VITAMINS C/D	☐ MULTIVITAMINS	☐ OTHER

WEEKLY CARE	MON	TUE	WED	THURS	FRI	SAT	SUN
REFRESH BOWL							
FRESH FOOD							
FRESH WATER							
PLAYING TIME							
EYE CHECK							
EAR CHECK							
CLAW CHECK							
TEETH CHECK							

NOTES & OBSERVATIONS

FINAL ASSESSMENT

OBJECTIVES MET	☐ YES	☐ NO	OWNER SIGNED:

RAT NAME		DATE
RAT SPECIES		
WEEK STARTING		/ /

RAT AGE			HEALTHCARE RATING	1	2	3	4	5
RAT GENDER								

TIME OF YEAR					FOOD GIVEN	

CLEAN START			PET RAT CARE CHECKLIST	
CLEAN END			☐ CLEAN CAGE	☐ CHANGE TOYS
RAT SIZE			☐ CHANGE LINING	☐ FOOD/WATER
RAT WEIGHT			☐ CHANGE STRAW	☐ CLEAN FAECES

RAT VITAMINS & SUPPLEMENTS

☐ VITAMIN A	☐ VITAMINS C/D	☐ MULTIVITAMINS	☐ OTHER

WEEKLY CARE	MON	TUE	WED	THURS	FRI	SAT	SUN
REFRESH BOWL							
FRESH FOOD							
FRESH WATER							
PLAYING TIME							
EYE CHECK							
EAR CHECK							
CLAW CHECK							
TEETH CHECK							

NOTES & OBSERVATIONS

FINAL ASSESSMENT

OBJECTIVES MET ☐ YES	☐ NO	OWNER SIGNED:	

RAT NAME		DATE
RAT SPECIES		
WEEK STARTING		/ /

RAT AGE			HEALTHCARE RATING	1	2	3	4	5
RAT GENDER								

TIME OF YEAR						FOOD GIVEN	

CLEAN START		
CLEAN END		
RAT SIZE		
RAT WEIGHT		

PET RAT CARE CHECKLIST

☐ CLEAN CAGE	☐ CHANGE TOYS
☐ CHANGE LINING	☐ FOOD/WATER
☐ CHANGE STRAW	☐ CLEAN FAECES

RAT VITAMINS & SUPPLEMENTS

☐ VITAMIN A	☐ VITAMINS C/D	☐ MULTIVITAMINS	☐ OTHER

WEEKLY CARE	MON	TUE	WED	THURS	FRI	SAT	SUN
REFRESH BOWL							
FRESH FOOD							
FRESH WATER							
PLAYING TIME							
EYE CHECK							
EAR CHECK							
CLAW CHECK							
TEETH CHECK							

NOTES & OBSERVATIONS

FINAL ASSESSMENT

OBJECTIVES MET	☐ YES	☐ NO	OWNER SIGNED:

RAT NAME		DATE
RAT SPECIES		
WEEK STARTING		/ /

RAT AGE			HEALTHCARE RATING	1	2	3	4	5
RAT GENDER								

TIME OF YEAR					FOOD GIVEN	

CLEAN START		
CLEAN END		
RAT SIZE		
RAT WEIGHT		

PET RAT CARE CHECKLIST

☐ CLEAN CAGE	☐ CHANGE TOYS
☐ CHANGE LINING	☐ FOOD/WATER
☐ CHANGE STRAW	☐ CLEAN FAECES

RAT VITAMINS & SUPPLEMENTS

☐ VITAMIN A	☐ VITAMINS C/D	☐ MULTIVITAMINS	☐ OTHER

WEEKLY CARE	MON	TUE	WED	THURS	FRI	SAT	SUN
REFRESH BOWL							
FRESH FOOD							
FRESH WATER							
PLAYING TIME							
EYE CHECK							
EAR CHECK							
CLAW CHECK							
TEETH CHECK							

NOTES & OBSERVATIONS

FINAL ASSESSMENT

OBJECTIVES MET	☐ YES	☐ NO	OWNER SIGNED:

RAT NAME		DATE
RAT SPECIES		
WEEK STARTING		/ /

RAT AGE		HEALTHCARE RATING	1	2	3	4	5
RAT GENDER							

TIME OF YEAR					FOOD GIVEN	

CLEAN START		
CLEAN END		
RAT SIZE		
RAT WEIGHT		

PET RAT CARE CHECKLIST

☐ CLEAN CAGE	☐ CHANGE TOYS
☐ CHANGE LINING	☐ FOOD/WATER
☐ CHANGE STRAW	☐ CLEAN FAECES

RAT VITAMINS & SUPPLEMENTS

☐ VITAMIN A	☐ VITAMINS C/D	☐ MULTIVITAMINS	☐ OTHER

WEEKLY CARE	MON	TUE	WED	THURS	FRI	SAT	SUN
REFRESH BOWL							
FRESH FOOD							
FRESH WATER							
PLAYING TIME							
EYE CHECK							
EAR CHECK							
CLAW CHECK							
TEETH CHECK							

NOTES & OBSERVATIONS

FINAL ASSESSMENT

OBJECTIVES MET	☐ YES	☐ NO	OWNER SIGNED:

RAT NAME		DATE
RAT SPECIES		
WEEK STARTING		/ /

RAT AGE		HEALTHCARE RATING	1	2	3	4	5
RAT GENDER							

TIME OF YEAR		FOOD GIVEN

CLEAN START		PET RAT CARE CHECKLIST	
CLEAN END		☐ CLEAN CAGE	☐ CHANGE TOYS
RAT SIZE		☐ CHANGE LINING	☐ FOOD/WATER
RAT WEIGHT		☐ CHANGE STRAW	☐ CLEAN FAECES

RAT VITAMINS & SUPPLEMENTS

☐ VITAMIN A ☐ VITAMINS C/D ☐ MULTIVITAMINS ☐ OTHER

WEEKLY CARE	MON	TUE	WED	THURS	FRI	SAT	SUN
REFRESH BOWL							
FRESH FOOD							
FRESH WATER							
PLAYING TIME							
EYE CHECK							
EAR CHECK							
CLAW CHECK							
TEETH CHECK							

NOTES & OBSERVATIONS

FINAL ASSESSMENT

OBJECTIVES MET ☐ YES ☐ NO OWNER SIGNED:

RAT NAME		DATE
RAT SPECIES		
WEEK STARTING		/ /

RAT AGE		HEALTHCARE RATING	1 2 3 4 5
RAT GENDER			
TIME OF YEAR		FOOD GIVEN	
CLEAN START			
CLEAN END			
RAT SIZE			
RAT WEIGHT			

PET RAT CARE CHECKLIST

☐ CLEAN CAGE	☐ CHANGE TOYS
☐ CHANGE LINING	☐ FOOD/WATER
☐ CHANGE STRAW	☐ CLEAN FAECES

RAT VITAMINS & SUPPLEMENTS

☐ VITAMIN A ☐ VITAMINS C/D ☐ MULTIVITAMINS ☐ OTHER

WEEKLY CARE	MON	TUE	WED	THURS	FRI	SAT	SUN
REFRESH BOWL							
FRESH FOOD							
FRESH WATER							
PLAYING TIME							
EYE CHECK							
EAR CHECK							
CLAW CHECK							
TEETH CHECK							

NOTES & OBSERVATIONS

FINAL ASSESSMENT

OBJECTIVES MET ☐ YES ☐ NO OWNER SIGNED:

RAT NAME		DATE
RAT SPECIES		
WEEK STARTING		/ /

RAT AGE		HEALTHCARE RATING	1	2	3	4	5
RAT GENDER							

TIME OF YEAR					FOOD GIVEN	

CLEAN START		
CLEAN END		
RAT SIZE		
RAT WEIGHT		

PET RAT CARE CHECKLIST

☐ CLEAN CAGE	☐ CHANGE TOYS
☐ CHANGE LINING	☐ FOOD/WATER
☐ CHANGE STRAW	☐ CLEAN FAECES

RAT VITAMINS & SUPPLEMENTS

☐ VITAMIN A	☐ VITAMINS C/D	☐ MULTIVITAMINS	☐ OTHER

WEEKLY CARE	MON	TUE	WED	THURS	FRI	SAT	SUN
REFRESH BOWL							
FRESH FOOD							
FRESH WATER							
PLAYING TIME							
EYE CHECK							
EAR CHECK							
CLAW CHECK							
TEETH CHECK							

NOTES & OBSERVATIONS

FINAL ASSESSMENT

OBJECTIVES MET ☐ YES	☐ NO	OWNER SIGNED:	

RAT NAME		DATE
RAT SPECIES		/ /
WEEK STARTING		

RAT AGE		HEALTHCARE RATING	1	2	3	4	5
RAT GENDER							

TIME OF YEAR					FOOD GIVEN	

CLEAN START		
CLEAN END		
RAT SIZE		
RAT WEIGHT		

PET RAT CARE CHECKLIST

☐ CLEAN CAGE	☐ CHANGE TOYS
☐ CHANGE LINING	☐ FOOD/WATER
☐ CHANGE STRAW	☐ CLEAN FAECES

RAT VITAMINS & SUPPLEMENTS

☐ VITAMIN A	☐ VITAMINS C/D	☐ MULTIVITAMINS	☐ OTHER

WEEKLY CARE	MON	TUE	WED	THURS	FRI	SAT	SUN
REFRESH BOWL							
FRESH FOOD							
FRESH WATER							
PLAYING TIME							
EYE CHECK							
EAR CHECK							
CLAW CHECK							
TEETH CHECK							

NOTES & OBSERVATIONS

FINAL ASSESSMENT

OBJECTIVES MET ☐ YES	☐ NO	OWNER SIGNED:	

RAT NAME		DATE
RAT SPECIES		
WEEK STARTING		/ /

RAT AGE		HEALTHCARE RATING	1	2	3	4	5
RAT GENDER							

TIME OF YEAR		FOOD GIVEN

CLEAN START		**PET RAT CARE CHECKLIST**	
CLEAN END		☐ CLEAN CAGE	☐ CHANGE TOYS
RAT SIZE		☐ CHANGE LINING	☐ FOOD/WATER
RAT WEIGHT		☐ CHANGE STRAW	☐ CLEAN FAECES

RAT VITAMINS & SUPPLEMENTS

☐ VITAMIN A ☐ VITAMINS C/D ☐ MULTIVITAMINS ☐ OTHER

WEEKLY CARE	MON	TUE	WED	THURS	FRI	SAT	SUN
REFRESH BOWL							
FRESH FOOD							
FRESH WATER							
PLAYING TIME							
EYE CHECK							
EAR CHECK							
CLAW CHECK							
TEETH CHECK							

NOTES & OBSERVATIONS

FINAL ASSESSMENT

OBJECTIVES MET ☐ YES ☐ NO OWNER SIGNED:

		DATE
RAT NAME		
RAT SPECIES		
WEEK STARTING		/ /

RAT AGE / RAT GENDER			HEALTHCARE RATING	1	2	3	4	5
TIME OF YEAR					FOOD GIVEN				
CLEAN START / CLEAN END									
RAT SIZE / RAT WEIGHT									

PET RAT CARE CHECKLIST

- ☐ CLEAN CAGE
- ☐ CHANGE TOYS
- ☐ CHANGE LINING
- ☐ FOOD/WATER
- ☐ CHANGE STRAW
- ☐ CLEAN FAECES

RAT VITAMINS & SUPPLEMENTS

☐ VITAMIN A ☐ VITAMINS C/D ☐ MULTIVITAMINS ☐ OTHER

WEEKLY CARE	MON	TUE	WED	THURS	FRI	SAT	SUN
REFRESH BOWL							
FRESH FOOD							
FRESH WATER							
PLAYING TIME							
EYE CHECK							
EAR CHECK							
CLAW CHECK							
TEETH CHECK							

NOTES & OBSERVATIONS

FINAL ASSESSMENT

OBJECTIVES MET ☐ YES ☐ NO OWNER SIGNED:

RAT NAME		DATE
RAT SPECIES		
WEEK STARTING		/ /

RAT AGE		HEALTHCARE RATING	1	2	3	4	5
RAT GENDER							

TIME OF YEAR					FOOD GIVEN	

CLEAN START	
CLEAN END	
RAT SIZE	
RAT WEIGHT	

PET RAT CARE CHECKLIST

☐ CLEAN CAGE	☐ CHANGE TOYS
☐ CHANGE LINING	☐ FOOD/WATER
☐ CHANGE STRAW	☐ CLEAN FAECES

RAT VITAMINS & SUPPLEMENTS

☐ VITAMIN A	☐ VITAMINS C/D	☐ MULTIVITAMINS	☐ OTHER

WEEKLY CARE	MON	TUE	WED	THURS	FRI	SAT	SUN
REFRESH BOWL							
FRESH FOOD							
FRESH WATER							
PLAYING TIME							
EYE CHECK							
EAR CHECK							
CLAW CHECK							
TEETH CHECK							

NOTES & OBSERVATIONS

FINAL ASSESSMENT

OBJECTIVES MET ☐ YES ☐ NO OWNER SIGNED:

RAT NAME		DATE
RAT SPECIES		
WEEK STARTING		/ /

RAT AGE			HEALTHCARE RATING	1	2	3	4	5
RAT GENDER								
TIME OF YEAR			FOOD GIVEN					

CLEAN START			PET RAT CARE CHECKLIST	
CLEAN END			☐ CLEAN CAGE	☐ CHANGE TOYS
RAT SIZE			☐ CHANGE LINING	☐ FOOD/WATER
RAT WEIGHT			☐ CHANGE STRAW	☐ CLEAN FAECES

RAT VITAMINS & SUPPLEMENTS

☐ VITAMIN A	☐ VITAMINS C/D	☐ MULTIVITAMINS	☐ OTHER

WEEKLY CARE	MON	TUE	WED	THURS	FRI	SAT	SUN
REFRESH BOWL							
FRESH FOOD							
FRESH WATER							
PLAYING TIME							
EYE CHECK							
EAR CHECK							
CLAW CHECK							
TEETH CHECK							

NOTES & OBSERVATIONS

FINAL ASSESSMENT

OBJECTIVES MET ☐ YES	☐ NO	OWNER SIGNED:	

RAT NAME		DATE
RAT SPECIES		
WEEK STARTING		/ /

RAT AGE		HEALTHCARE RATING	1	2	3	4	5
RAT GENDER							

TIME OF YEAR		FOOD GIVEN	
CLEAN START		PET RAT CARE CHECKLIST	
CLEAN END		☐ CLEAN CAGE	☐ CHANGE TOYS
RAT SIZE		☐ CHANGE LINING	☐ FOOD/WATER
RAT WEIGHT		☐ CHANGE STRAW	☐ CLEAN FAECES

RAT VITAMINS & SUPPLEMENTS

☐ VITAMIN A	☐ VITAMINS C/D	☐ MULTIVITAMINS	☐ OTHER

WEEKLY CARE	MON	TUE	WED	THURS	FRI	SAT	SUN
REFRESH BOWL							
FRESH FOOD							
FRESH WATER							
PLAYING TIME							
EYE CHECK							
EAR CHECK							
CLAW CHECK							
TEETH CHECK							

NOTES & OBSERVATIONS

FINAL ASSESSMENT

OBJECTIVES MET ☐ YES ☐ NO OWNER SIGNED:

RAT NAME		DATE
RAT SPECIES		
WEEK STARTING		/ /

RAT AGE			HEALTHCARE RATING	1	2	3	4	5
RAT GENDER								

TIME OF YEAR						FOOD GIVEN	

CLEAN START			
CLEAN END			
RAT SIZE			
RAT WEIGHT			

PET RAT CARE CHECKLIST

☐ CLEAN CAGE	☐ CHANGE TOYS
☐ CHANGE LINING	☐ FOOD/WATER
☐ CHANGE STRAW	☐ CLEAN FAECES

RAT VITAMINS & SUPPLEMENTS

☐ VITAMIN A	☐ VITAMINS C/D	☐ MULTIVITAMINS	☐ OTHER

WEEKLY CARE	MON	TUE	WED	THURS	FRI	SAT	SUN
REFRESH BOWL							
FRESH FOOD							
FRESH WATER							
PLAYING TIME							
EYE CHECK							
EAR CHECK							
CLAW CHECK							
TEETH CHECK							

NOTES & OBSERVATIONS

FINAL ASSESSMENT

OBJECTIVES MET ☐ YES ☐ NO OWNER SIGNED:

RAT NAME		DATE
RAT SPECIES		
WEEK STARTING		/ /

RAT AGE			HEALTHCARE RATING	1	2	3	4	5
RAT GENDER								

TIME OF YEAR					FOOD GIVEN	

CLEAN START		
CLEAN END		
RAT SIZE		
RAT WEIGHT		

PET RAT CARE CHECKLIST	
☐ CLEAN CAGE	☐ CHANGE TOYS
☐ CHANGE LINING	☐ FOOD/WATER
☐ CHANGE STRAW	☐ CLEAN FAECES

RAT VITAMINS & SUPPLEMENTS

☐ VITAMIN A ☐ VITAMINS C/D ☐ MULTIVITAMINS ☐ OTHER

WEEKLY CARE	MON	TUE	WED	THURS	FRI	SAT	SUN
REFRESH BOWL							
FRESH FOOD							
FRESH WATER							
PLAYING TIME							
EYE CHECK							
EAR CHECK							
CLAW CHECK							
TEETH CHECK							

NOTES & OBSERVATIONS

FINAL ASSESSMENT

OBJECTIVES MET ☐ YES ☐ NO OWNER SIGNED:

RAT NAME		DATE
RAT SPECIES		
WEEK STARTING		/ /

RAT AGE		HEALTHCARE RATING	1 2 3 4 5
RAT GENDER			
TIME OF YEAR		FOOD GIVEN	
CLEAN START			
CLEAN END			
RAT SIZE			
RAT WEIGHT			

PET RAT CARE CHECKLIST

☐ CLEAN CAGE	☐ CHANGE TOYS
☐ CHANGE LINING	☐ FOOD/WATER
☐ CHANGE STRAW	☐ CLEAN FAECES

RAT VITAMINS & SUPPLEMENTS

☐ VITAMIN A ☐ VITAMINS C/D ☐ MULTIVITAMINS ☐ OTHER

WEEKLY CARE	MON	TUE	WED	THURS	FRI	SAT	SUN
REFRESH BOWL							
FRESH FOOD							
FRESH WATER							
PLAYING TIME							
EYE CHECK							
EAR CHECK							
CLAW CHECK							
TEETH CHECK							

NOTES & OBSERVATIONS

FINAL ASSESSMENT

OBJECTIVES MET ☐ YES ☐ NO OWNER SIGNED:

RAT NAME		DATE
RAT SPECIES		
WEEK STARTING		/ /

RAT AGE		HEALTHCARE RATING	1	2	3	4	5
RAT GENDER							
TIME OF YEAR		FOOD GIVEN					
CLEAN START		PET RAT CARE CHECKLIST					
CLEAN END		☐ CLEAN CAGE	☐ CHANGE TOYS				
RAT SIZE		☐ CHANGE LINING	☐ FOOD/WATER				
RAT WEIGHT		☐ CHANGE STRAW	☐ CLEAN FAECES				

RAT VITAMINS & SUPPLEMENTS

☐ VITAMIN A ☐ VITAMINS C/D ☐ MULTIVITAMINS ☐ OTHER

WEEKLY CARE	MON	TUE	WED	THURS	FRI	SAT	SUN
REFRESH BOWL							
FRESH FOOD							
FRESH WATER							
PLAYING TIME							
EYE CHECK							
EAR CHECK							
CLAW CHECK							
TEETH CHECK							

NOTES & OBSERVATIONS

FINAL ASSESSMENT

OBJECTIVES MET ☐ YES ☐ NO OWNER SIGNED:

RAT NAME		DATE
RAT SPECIES		
WEEK STARTING		/ /

RAT AGE		HEALTHCARE RATING	1	2	3	4	5
RAT GENDER							

TIME OF YEAR		FOOD GIVEN	

CLEAN START	
CLEAN END	
RAT SIZE	
RAT WEIGHT	

PET RAT CARE CHECKLIST

☐ CLEAN CAGE	☐ CHANGE TOYS
☐ CHANGE LINING	☐ FOOD/WATER
☐ CHANGE STRAW	☐ CLEAN FAECES

RAT VITAMINS & SUPPLEMENTS

☐ VITAMIN A	☐ VITAMINS C/D	☐ MULTIVITAMINS	☐ OTHER

WEEKLY CARE	MON	TUE	WED	THURS	FRI	SAT	SUN
REFRESH BOWL							
FRESH FOOD							
FRESH WATER							
PLAYING TIME							
EYE CHECK							
EAR CHECK							
CLAW CHECK							
TEETH CHECK							

NOTES & OBSERVATIONS

FINAL ASSESSMENT

OBJECTIVES MET ☐ YES ☐ NO OWNER SIGNED:

RAT NAME		DATE
RAT SPECIES		
WEEK STARTING		/ /

RAT AGE		HEALTHCARE RATING	1	2	3	4	5
RAT GENDER							

TIME OF YEAR		FOOD GIVEN

CLEAN START		PET RAT CARE CHECKLIST	
CLEAN END		☐ CLEAN CAGE	☐ CHANGE TOYS
RAT SIZE		☐ CHANGE LINING	☐ FOOD/WATER
RAT WEIGHT		☐ CHANGE STRAW	☐ CLEAN FAECES

RAT VITAMINS & SUPPLEMENTS

☐ VITAMIN A ☐ VITAMINS C/D ☐ MULTIVITAMINS ☐ OTHER

WEEKLY CARE	MON	TUE	WED	THURS	FRI	SAT	SUN
REFRESH BOWL							
FRESH FOOD							
FRESH WATER							
PLAYING TIME							
EYE CHECK							
EAR CHECK							
CLAW CHECK							
TEETH CHECK							

NOTES & OBSERVATIONS

FINAL ASSESSMENT

OBJECTIVES MET ☐ YES ☐ NO OWNER SIGNED:

RAT NAME		DATE
RAT SPECIES		
WEEK STARTING		/ /

RAT AGE		HEALTHCARE RATING	1 2 3 4 5
RAT GENDER			
TIME OF YEAR		FOOD GIVEN	
CLEAN START			
CLEAN END			
RAT SIZE			
RAT WEIGHT			

PET RAT CARE CHECKLIST

☐ CLEAN CAGE	☐ CHANGE TOYS
☐ CHANGE LINING	☐ FOOD/WATER
☐ CHANGE STRAW	☐ CLEAN FAECES

RAT VITAMINS & SUPPLEMENTS

☐ VITAMIN A	☐ VITAMINS C/D	☐ MULTIVITAMINS	☐ OTHER

WEEKLY CARE	MON	TUE	WED	THURS	FRI	SAT	SUN
REFRESH BOWL							
FRESH FOOD							
FRESH WATER							
PLAYING TIME							
EYE CHECK							
EAR CHECK							
CLAW CHECK							
TEETH CHECK							

NOTES & OBSERVATIONS

FINAL ASSESSMENT

OBJECTIVES MET ☐ YES ☐ NO OWNER SIGNED:

RAT NAME		DATE
RAT SPECIES		
WEEK STARTING		/ /

RAT AGE			HEALTHCARE RATING	1	2	3	4	5
RAT GENDER								

TIME OF YEAR					FOOD GIVEN

CLEAN START		
CLEAN END		
RAT SIZE		
RAT WEIGHT		

PET RAT CARE CHECKLIST

☐ CLEAN CAGE	☐ CHANGE TOYS
☐ CHANGE LINING	☐ FOOD/WATER
☐ CHANGE STRAW	☐ CLEAN FAECES

RAT VITAMINS & SUPPLEMENTS

☐ VITAMIN A ☐ VITAMINS C/D ☐ MULTIVITAMINS ☐ OTHER

WEEKLY CARE	MON	TUE	WED	THURS	FRI	SAT	SUN
REFRESH BOWL							
FRESH FOOD							
FRESH WATER							
PLAYING TIME							
EYE CHECK							
EAR CHECK							
CLAW CHECK							
TEETH CHECK							

NOTES & OBSERVATIONS

FINAL ASSESSMENT

OBJECTIVES MET ☐ YES ☐ NO OWNER SIGNED:

RAT NAME		DATE
RAT SPECIES		
WEEK STARTING		/ /

RAT AGE		
RAT GENDER		

HEALTHCARE RATING	1	2	3	4	5

TIME OF YEAR	
FOOD GIVEN	
CLEAN START	
CLEAN END	
RAT SIZE	
RAT WEIGHT	

PET RAT CARE CHECKLIST

☐ CLEAN CAGE	☐ CHANGE TOYS
☐ CHANGE LINING	☐ FOOD/WATER
☐ CHANGE STRAW	☐ CLEAN FAECES

RAT VITAMINS & SUPPLEMENTS

☐ VITAMIN A	☐ VITAMINS C/D	☐ MULTIVITAMINS	☐ OTHER

WEEKLY CARE	MON	TUE	WED	THURS	FRI	SAT	SUN
REFRESH BOWL							
FRESH FOOD							
FRESH WATER							
PLAYING TIME							
EYE CHECK							
EAR CHECK							
CLAW CHECK							
TEETH CHECK							

NOTES & OBSERVATIONS

FINAL ASSESSMENT

OBJECTIVES MET	☐ YES	☐ NO	OWNER SIGNED:

RAT NAME		DATE
RAT SPECIES		
WEEK STARTING		/ /

RAT AGE		HEALTHCARE RATING	1	2	3	4	5
RAT GENDER							

TIME OF YEAR					FOOD GIVEN	

CLEAN START	
CLEAN END	

PET RAT CARE CHECKLIST

☐ CLEAN CAGE	☐ CHANGE TOYS
☐ CHANGE LINING	☐ FOOD/WATER
☐ CHANGE STRAW	☐ CLEAN FAECES

RAT SIZE	
RAT WEIGHT	

RAT VITAMINS & SUPPLEMENTS

☐ VITAMIN A ☐ VITAMINS C/D ☐ MULTIVITAMINS ☐ OTHER

WEEKLY CARE	MON	TUE	WED	THURS	FRI	SAT	SUN
REFRESH BOWL							
FRESH FOOD							
FRESH WATER							
PLAYING TIME							
EYE CHECK							
EAR CHECK							
CLAW CHECK							
TEETH CHECK							

NOTES & OBSERVATIONS

FINAL ASSESSMENT

OBJECTIVES MET ☐ YES ☐ NO OWNER SIGNED:

RAT NAME		DATE
RAT SPECIES		
WEEK STARTING		/ /

RAT AGE			HEALTHCARE RATING	1	2	3	4	5
RAT GENDER								
TIME OF YEAR			FOOD GIVEN					
CLEAN START								
CLEAN END								
RAT SIZE								
RAT WEIGHT								

PET RAT CARE CHECKLIST

- ☐ CLEAN CAGE
- ☐ CHANGE TOYS
- ☐ CHANGE LINING
- ☐ FOOD/WATER
- ☐ CHANGE STRAW
- ☐ CLEAN FAECES

RAT VITAMINS & SUPPLEMENTS

☐ VITAMIN A ☐ VITAMINS C/D ☐ MULTIVITAMINS ☐ OTHER

WEEKLY CARE	MON	TUE	WED	THURS	FRI	SAT	SUN
REFRESH BOWL							
FRESH FOOD							
FRESH WATER							
PLAYING TIME							
EYE CHECK							
EAR CHECK							
CLAW CHECK							
TEETH CHECK							

NOTES & OBSERVATIONS

FINAL ASSESSMENT

OBJECTIVES MET ☐ YES ☐ NO OWNER SIGNED:

RAT NAME		DATE
RAT SPECIES		
WEEK STARTING		/ /

RAT AGE		HEALTHCARE RATING	1	2	3	4	5
RAT GENDER							

TIME OF YEAR		FOOD GIVEN	
CLEAN START		PET RAT CARE CHECKLIST	
CLEAN END		☐ CLEAN CAGE	☐ CHANGE TOYS
RAT SIZE		☐ CHANGE LINING	☐ FOOD/WATER
RAT WEIGHT		☐ CHANGE STRAW	☐ CLEAN FAECES

RAT VITAMINS & SUPPLEMENTS

☐ VITAMIN A ☐ VITAMINS C/D ☐ MULTIVITAMINS ☐ OTHER

WEEKLY CARE	MON	TUE	WED	THURS	FRI	SAT	SUN
REFRESH BOWL							
FRESH FOOD							
FRESH WATER							
PLAYING TIME							
EYE CHECK							
EAR CHECK							
CLAW CHECK							
TEETH CHECK							

NOTES & OBSERVATIONS

FINAL ASSESSMENT

OBJECTIVES MET ☐ YES ☐ NO OWNER SIGNED:

RAT NAME		DATE
RAT SPECIES		
WEEK STARTING		/ /

RAT AGE		HEALTHCARE RATING	1 2 3 4 5
RAT GENDER			
TIME OF YEAR		FOOD GIVEN	
CLEAN START		PET RAT CARE CHECKLIST	
CLEAN END		☐ CLEAN CAGE	☐ CHANGE TOYS
RAT SIZE		☐ CHANGE LINING	☐ FOOD/WATER
RAT WEIGHT		☐ CHANGE STRAW	☐ CLEAN FAECES

RAT VITAMINS & SUPPLEMENTS

☐ VITAMIN A ☐ VITAMINS C/D ☐ MULTIVITAMINS ☐ OTHER

WEEKLY CARE	MON	TUE	WED	THURS	FRI	SAT	SUN
REFRESH BOWL							
FRESH FOOD							
FRESH WATER							
PLAYING TIME							
EYE CHECK							
EAR CHECK							
CLAW CHECK							
TEETH CHECK							

NOTES & OBSERVATIONS

FINAL ASSESSMENT

OBJECTIVES MET ☐ YES ☐ NO OWNER SIGNED:

RAT NAME		DATE
RAT SPECIES		
WEEK STARTING		/ /

RAT AGE		HEALTHCARE RATING	1	2	3	4	5
RAT GENDER							

TIME OF YEAR		FOOD GIVEN	

CLEAN START		PET RAT CARE CHECKLIST	
CLEAN END		☐ CLEAN CAGE	☐ CHANGE TOYS
RAT SIZE		☐ CHANGE LINING	☐ FOOD/WATER
RAT WEIGHT		☐ CHANGE STRAW	☐ CLEAN FAECES

RAT VITAMINS & SUPPLEMENTS

☐ VITAMIN A	☐ VITAMINS C/D	☐ MULTIVITAMINS	☐ OTHER

WEEKLY CARE	MON	TUE	WED	THURS	FRI	SAT	SUN
REFRESH BOWL							
FRESH FOOD							
FRESH WATER							
PLAYING TIME							
EYE CHECK							
EAR CHECK							
CLAW CHECK							
TEETH CHECK							

NOTES & OBSERVATIONS

FINAL ASSESSMENT

OBJECTIVES MET ☐ YES	☐ NO	OWNER SIGNED:

RAT NAME		DATE
RAT SPECIES		
WEEK STARTING		/ /

RAT AGE		HEALTHCARE RATING	1	2	3	4	5
RAT GENDER							

TIME OF YEAR		FOOD GIVEN	

CLEAN START		**PET RAT CARE CHECKLIST**	
CLEAN END		☐ CLEAN CAGE	☐ CHANGE TOYS
RAT SIZE		☐ CHANGE LINING	☐ FOOD/WATER
RAT WEIGHT		☐ CHANGE STRAW	☐ CLEAN FAECES

RAT VITAMINS & SUPPLEMENTS

☐ VITAMIN A ☐ VITAMINS C/D ☐ MULTIVITAMINS ☐ OTHER

WEEKLY CARE	MON	TUE	WED	THURS	FRI	SAT	SUN
REFRESH BOWL							
FRESH FOOD							
FRESH WATER							
PLAYING TIME							
EYE CHECK							
EAR CHECK							
CLAW CHECK							
TEETH CHECK							

NOTES & OBSERVATIONS

FINAL ASSESSMENT

OBJECTIVES MET ☐ YES ☐ NO OWNER SIGNED:

RAT NAME		DATE
RAT SPECIES		
WEEK STARTING		/ /

RAT AGE			HEALTHCARE RATING	1	2	3	4	5
RAT GENDER								
TIME OF YEAR			FOOD GIVEN					
CLEAN START								
CLEAN END								
RAT SIZE								
RAT WEIGHT								

PET RAT CARE CHECKLIST

☐ CLEAN CAGE	☐ CHANGE TOYS
☐ CHANGE LINING	☐ FOOD/WATER
☐ CHANGE STRAW	☐ CLEAN FAECES

RAT VITAMINS & SUPPLEMENTS

☐ VITAMIN A	☐ VITAMINS C/D	☐ MULTIVITAMINS	☐ OTHER

WEEKLY CARE	MON	TUE	WED	THURS	FRI	SAT	SUN
REFRESH BOWL							
FRESH FOOD							
FRESH WATER							
PLAYING TIME							
EYE CHECK							
EAR CHECK							
CLAW CHECK							
TEETH CHECK							

NOTES & OBSERVATIONS

FINAL ASSESSMENT

OBJECTIVES MET	☐ YES	☐ NO	OWNER SIGNED:

RAT NAME		DATE
RAT SPECIES		/ /
WEEK STARTING		

RAT AGE			HEALTHCARE RATING	1	2	3	4	5
RAT GENDER								
TIME OF YEAR			FOOD GIVEN					

CLEAN START		
CLEAN END		
RAT SIZE		
RAT WEIGHT		

PET RAT CARE CHECKLIST

- ☐ CLEAN CAGE
- ☐ CHANGE TOYS
- ☐ CHANGE LINING
- ☐ FOOD/WATER
- ☐ CHANGE STRAW
- ☐ CLEAN FAECES

RAT VITAMINS & SUPPLEMENTS

☐ VITAMIN A ☐ VITAMINS C/D ☐ MULTIVITAMINS ☐ OTHER

WEEKLY CARE	MON	TUE	WED	THURS	FRI	SAT	SUN
REFRESH BOWL							
FRESH FOOD							
FRESH WATER							
PLAYING TIME							
EYE CHECK							
EAR CHECK							
CLAW CHECK							
TEETH CHECK							

NOTES & OBSERVATIONS

FINAL ASSESSMENT

OBJECTIVES MET ☐ YES ☐ NO OWNER SIGNED:

RAT NAME		DATE
RAT SPECIES		
WEEK STARTING		/ /

RAT AGE		HEALTHCARE RATING	1	2	3	4	5
RAT GENDER							

TIME OF YEAR		FOOD GIVEN

CLEAN START	
CLEAN END	
RAT SIZE	
RAT WEIGHT	

PET RAT CARE CHECKLIST

- ☐ CLEAN CAGE
- ☐ CHANGE TOYS
- ☐ CHANGE LINING
- ☐ FOOD/WATER
- ☐ CHANGE STRAW
- ☐ CLEAN FAECES

RAT VITAMINS & SUPPLEMENTS

☐ VITAMIN A ☐ VITAMINS C/D ☐ MULTIVITAMINS ☐ OTHER

WEEKLY CARE	MON	TUE	WED	THURS	FRI	SAT	SUN
REFRESH BOWL							
FRESH FOOD							
FRESH WATER							
PLAYING TIME							
EYE CHECK							
EAR CHECK							
CLAW CHECK							
TEETH CHECK							

NOTES & OBSERVATIONS

FINAL ASSESSMENT

OBJECTIVES MET ☐ YES ☐ NO OWNER SIGNED:

RAT NAME		DATE
RAT SPECIES		/ /
WEEK STARTING		

RAT AGE		HEALTHCARE RATING	1	2	3	4	5
RAT GENDER							

TIME OF YEAR		FOOD GIVEN	
CLEAN START			
CLEAN END			
RAT SIZE			
RAT WEIGHT			

PET RAT CARE CHECKLIST

☐ CLEAN CAGE	☐ CHANGE TOYS
☐ CHANGE LINING	☐ FOOD/WATER
☐ CHANGE STRAW	☐ CLEAN FAECES

RAT VITAMINS & SUPPLEMENTS

☐ VITAMIN A ☐ VITAMINS C/D ☐ MULTIVITAMINS ☐ OTHER

WEEKLY CARE	MON	TUE	WED	THURS	FRI	SAT	SUN
REFRESH BOWL							
FRESH FOOD							
FRESH WATER							
PLAYING TIME							
EYE CHECK							
EAR CHECK							
CLAW CHECK							
TEETH CHECK							

NOTES & OBSERVATIONS

FINAL ASSESSMENT

OBJECTIVES MET ☐ YES ☐ NO OWNER SIGNED:

		DATE
RAT NAME		
RAT SPECIES		
WEEK STARTING		/ /

RAT AGE			HEALTHCARE RATING	1	2	3	4	5
RAT GENDER								
TIME OF YEAR			FOOD GIVEN					
CLEAN START			PET RAT CARE CHECKLIST					
CLEAN END			☐ CLEAN CAGE	☐ CHANGE TOYS				
RAT SIZE			☐ CHANGE LINING	☐ FOOD/WATER				
RAT WEIGHT			☐ CHANGE STRAW	☐ CLEAN FAECES				

RAT VITAMINS & SUPPLEMENTS

☐ VITAMIN A ☐ VITAMINS C/D ☐ MULTIVITAMINS ☐ OTHER

WEEKLY CARE	MON	TUE	WED	THURS	FRI	SAT	SUN
REFRESH BOWL							
FRESH FOOD							
FRESH WATER							
PLAYING TIME							
EYE CHECK							
EAR CHECK							
CLAW CHECK							
TEETH CHECK							

NOTES & OBSERVATIONS

FINAL ASSESSMENT

OBJECTIVES MET ☐ YES ☐ NO OWNER SIGNED:

RAT NAME		DATE
RAT SPECIES		
WEEK STARTING		/ /

RAT AGE			HEALTHCARE RATING	1	2	3	4	5
RAT GENDER								

TIME OF YEAR						FOOD GIVEN	

CLEAN START		
CLEAN END		
RAT SIZE		
RAT WEIGHT		

PET RAT CARE CHECKLIST

☐ CLEAN CAGE	☐ CHANGE TOYS
☐ CHANGE LINING	☐ FOOD/WATER
☐ CHANGE STRAW	☐ CLEAN FAECES

RAT VITAMINS & SUPPLEMENTS

☐ VITAMIN A ☐ VITAMINS C/D ☐ MULTIVITAMINS ☐ OTHER

WEEKLY CARE	MON	TUE	WED	THURS	FRI	SAT	SUN
REFRESH BOWL							
FRESH FOOD							
FRESH WATER							
PLAYING TIME							
EYE CHECK							
EAR CHECK							
CLAW CHECK							
TEETH CHECK							

NOTES & OBSERVATIONS

FINAL ASSESSMENT

OBJECTIVES MET ☐ YES ☐ NO OWNER SIGNED:

RAT NAME		DATE
RAT SPECIES		
WEEK STARTING		/ /

RAT AGE		HEALTHCARE RATING	1	2	3	4	5
RAT GENDER							

TIME OF YEAR		FOOD GIVEN

CLEAN START		**PET RAT CARE CHECKLIST**
CLEAN END		☐ CLEAN CAGE ☐ CHANGE TOYS
RAT SIZE		☐ CHANGE LINING ☐ FOOD/WATER
RAT WEIGHT		☐ CHANGE STRAW ☐ CLEAN FAECES

RAT VITAMINS & SUPPLEMENTS

☐ VITAMIN A ☐ VITAMINS C/D ☐ MULTIVITAMINS ☐ OTHER

WEEKLY CARE	MON	TUE	WED	THURS	FRI	SAT	SUN
REFRESH BOWL							
FRESH FOOD							
FRESH WATER							
PLAYING TIME							
EYE CHECK							
EAR CHECK							
CLAW CHECK							
TEETH CHECK							

NOTES & OBSERVATIONS

FINAL ASSESSMENT

OBJECTIVES MET ☐ YES ☐ NO OWNER SIGNED:

RAT NAME		DATE
RAT SPECIES		
WEEK STARTING		/ /

RAT AGE			HEALTHCARE RATING	1	2	3	4	5
RAT GENDER								
TIME OF YEAR			FOOD GIVEN					
CLEAN START			PET RAT CARE CHECKLIST					
CLEAN END			☐ CLEAN CAGE	☐ CHANGE TOYS				
RAT SIZE			☐ CHANGE LINING	☐ FOOD/WATER				
RAT WEIGHT			☐ CHANGE STRAW	☐ CLEAN FAECES				

RAT VITAMINS & SUPPLEMENTS

☐ VITAMIN A ☐ VITAMINS C/D ☐ MULTIVITAMINS ☐ OTHER

WEEKLY CARE	MON	TUE	WED	THURS	FRI	SAT	SUN
REFRESH BOWL							
FRESH FOOD							
FRESH WATER							
PLAYING TIME							
EYE CHECK							
EAR CHECK							
CLAW CHECK							
TEETH CHECK							

NOTES & OBSERVATIONS

FINAL ASSESSMENT

OBJECTIVES MET ☐ YES ☐ NO OWNER SIGNED:

Thanks For Reading!

Just a quick message to thank you so much for picking up one of our books! Our sincere hope is that this book has given you the value we always look to provide, and hope we can continue to produce quality books that will in anyway contribute to a better quality of life for our readers.

We are a small independent publisher based in London, UK and we work with talented authors from around the world, who dedicate every ounce of their effort to craft these memorable books for your reading pleasure.

The author of this title would love to hear about your experience with the book, and your review will go a long way to provide them with the insight and encouragement they need to keep creating the kind of books you want to read.

Your Opinion Makes a Real Difference.

If you want to let us know what you thought about the book, please visit the Amazon website and give us your review. We read every single review, no matter how long or short!

HAPPY READING!

MOONPEAK™

PUBLISHERS

Made in the USA
Monee, IL
04 November 2026

67379681R00069